Planning, Proposing, and Presenting Science Effectively

Second Edition

This concise guide to planning, writing, and presenting research, especially in biology and behavioral ecology, is intended for students at all levels. The guidelines apply equally to independent projects for introductory courses, directed-study projects, and undergraduate senior theses, as well as to master's theses, doctoral dissertations, and research aimed at publication.

We have updated several topics in this edition, most of which reflect technological advances that have changed the way science is proposed and presented and have opened up new ethical challenges.

New features of this edition include:

- Tips on the process of preparing grants and manuscripts for electronic submissions.
- Description of how to prepare effective presentations in PowerPoint®.
- Discussion of how to produce computer-generated posters.
- Extended comments on ethics, and a new appendix on ethical considerations.

This edition also continues to:

- Guide the reader through a discussion of the nature of scientific research, how to plan research, and how to obtain funding.
- Discuss writing a research proposal, whether for a formal proposal for thesis research to be written by a graduate student or for a research proposal for a funding agency such as the National Science Foundation (using the Doctoral Dissertation Improvement Grant format as a specific example).
- Deal with writing a research report such as a graduate thesis or a manuscript for publication in a research journal.

- Give advice and guidelines for presenting the results of research at research seminars and scientific meetings, and also provide useful tips on preparing abstracts and posters for scientific meetings.
- Show how to write an effective c.v.
- Give tips on how to write clearly, common abbreviations (including Latin phrases), and difficult inflections, as well as other issues.

Throughout, the book is illuminated with personal examples from the authors' own experiences with research in behavioral ecology, and there is an emphasis on problems associated with field studies.

All biologists and many others will find this a valuable resource and guide for the early years of their scientific careers. Established faculty will find it an essential instructional tool.

Planning, Proposing, and Presenting Science Effectively

Second Edition

A Guide for Graduate Students and Researchers in the Behavioral Sciences and Biology

JACK P. HAILMAN

KAREN B. STRIER

University of Wisconsin-Madison

CAMBRIDGE
UNIVERSITY PRESS

CAMBRIDGE UNIVERSITY PRESS
Cambridge, New York, Melbourne, Madrid, Cape Town, Singapore, São Paulo

CAMBRIDGE UNIVERSITY PRESS
The Edinburgh Building, Cambridge CB2 2RU, UK

Published in the United States of America by Cambridge University Press,
New York

www.cambridge.org
Information on this title: www.cambridge.org/9780521826471

First edition © Cambridge University Press 1997

Second edition © J. P. Hailman and K. B. Strier 2006

First published 1997

Second edition 2006

Printed in the United Kingdom at the University Press, Cambridge

A catalog record for this publication is available from the British Library

ISBN-13 978-0-521-82647-1 hardback
ISBN-10 0-521-82647-0 hardback
ISBN-13 978-0-521-533881 paperback
ISBN-10 0-521-53388-0 paperback

Contents

Preface

Training in the biological sciences, and indeed in most sciences, appropriately emphasizes mastery of techniques and theory – essential elements of any original research endeavor. The less glamorous steps to successful research are often neglected in the formal training programs of scientists – such as how to plan a study adequately, secure research funds effectively, present results (written and oral) clearly and interestingly, and present qualifications in a résumé. This book is a map to those activities, with examples drawn primarily from the authors' discipline of behavioral ecology. In the first edition, we integrated our own experiences with the most frequent problems encountered by our students to produce what we hoped would be a helpful, concise guide to the more practical aspects of scientific research. In this second edition, we have updated these tips to accompany the changes that advances in computer technology and the Internet have brought to the practical sides of submitting research grants and manuscripts and presenting research in talks and posters. We have also deleted one of our original appendixes, which provided addresses for funding agencies, because these are now readily available on the Web, and have replaced it with an appendix on ethics considerations. The basics of planning scientific research remain the same, however.

Our concern is with widely applicable skills. If you want to know details – such as what research questions are currently considered "hot" in your field, what organizations are currently offering monies for what kinds of research, what laboratory and field techniques are state-of-the-art, or how to use modern computer-intensive statistical applications properly – you must seek specific sources appropriate to your needs. In this book, you will find suggestions for how to develop your research plans and communicate the results of your research.

Any well-constructed book is aimed at a defined audience but is appropriate for others in proportion to their nearness to the target. The bull's-eye of our target is the master's or doctoral student in behavioral ecology, ethology, or comparative psychology. Going in one direction, senior thesis students and any undergraduate otherwise involved in research will find applicable material. In the other direction, postdoctoral researchers, untenured faculty, and even senior researchers should find useful information. In an entirely different dimension, related fields of study involving whole organisms – say, ecology, conservation biology, neuroethology, bioacoustics, and such – will be very near to the target. As one generalizes to other areas of biology and allied disciplines, the material will diminish somewhat in specific applicability but remain useful in generalities.

We realize that some researchers may seek different emphases: perhaps extensive treatment of how to turn the germ of an idea into a viable project, or many specific examples of funded research applications, or advanced training in how to "wow" an audience with an oral presentation. In our decisions concerning the balance and depth of topics covered, we recognized that no volume will fully satisfy everyone. There seems never to have been written a book that covers all the ground of this one, but there are several fine volumes that treat certain issues in greater detail than here. For this reason, and because we strived to keep this guide concise and handy, we have provided an annotated bibliography of works for reading and reference. If there is an appropriate work that escaped our list, the omission was unintentional.

ACKNOWLEDGMENTS
Every possible way of thanking others in a non-perfunctory manner seems already to have been used by some author somewhere. Our lack of originality therefore betrays no lack of appreciation in

acknowledging our students for their encouragement to produce this volume and their feedback on an earlier version, and our colleagues Robert L. Jeanne and Charles T. Snowdon for their extensive and thorough suggestions upon reading the entire manuscript of the first edition. We are also grateful to editor Robin Smith and our anonymous reviewers for their encouragement and helpful suggestions; to Fran Bartlett and Jeanne Borczuk of G & H Soho for the high quality of work and interactive spirit in copy editing, preparing proof, and indexing; and especially to Liz Hailman for all her help during the final months of producing the first edition. The second edition would not have emerged without the encouragement and support of senior editor Tracey Sanderson. This edition also incorporates the many suggestions Jeremy J. Hatch provided after the first edition was published, tips from Luisa Arnedo on generating posters using modern technology, and indirect input from other students and colleagues who have generously shared their experiences of planning, proposing, and presenting science.

Jupiter, Florida
J. P. H.

Madison, Wisconsin
K. B. S.

I   How to plan research

The great tragedy of Science – the slaying of a beautiful hypothesis by an ugly fact.

Thomas Henry Huxley (1825–95)

Success in science, as in most complex endeavors, depends partly on preparedness and planning. The three Persian princes of Serendip notwithstanding, a great truth is found in the aphorism that chance favors the prepared mind. No mere chapter could constitute a complete guide to planning research. This one attempts to cover the main points common to most projects. We begin with a background sketch of epistemology: how science as a whole works and the roles of individual investigators. That introduction provides a framework for discussing specific issues of planning research. Subsequent chapters deal with some of the major steps of doing science (e.g. how to write a grant proposal and how to communicate the results of research) that ensue after good planning.

SCIENTIFIC EPISTEMOLOGY

In order to plan research effectively, the investigator should understand how his or her activities fit into the endeavor of science as a whole. Some explanations of the "scientific method" confound epistemology – how we accumulate knowledge and understanding through science – with specific research activities of the individual investigator. This section attempts to disentangle the two by sketching the "big picture" first and then showing where the practicing scientist fits in.

Science as process and product

One can conceive of science as a cycle of activities and results based on procedures that are often referred to as the "hypothetico-deductive method." This method evaluates a hypothesis (more generally, a model) of how the world might work and deduces consequences that must be true if the world does actually work in the way posited. One can then check the real world to see whether these predicted consequences are verifiably true. The process is cyclical because if

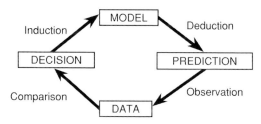

FIGURE 1.1 A schematic representation of the epistemological cycle

the consequences cannot be verified, then a new hypothesis (model) must be tried. Even if the predicted consequences can be verified, however, that result might be coincidence. Therefore, one strives to deduce new predictions from the same hypothesis and test these as well. The cycle, which we call here the epistemological cycle, can be represented schematically as in Figure 1.1.

Figure 1.1 uses some simple conventions of illustration. Boxes, which are labeled in FULL CAPS, represent statements that can be written down: products of the processes. Arrows, which are labeled in lower case, represent processes that yield the statements in the boxes to which they point. Furthermore, each arrow originates at a statement and points to another statement, so as to show the sequence of steps in the process of science. As the diagram represents a cycle, we can scrutinize its parts beginning at any arbitrary place and then return to that place by completing the cycle. The practicing scientist, however, usually enters the cycle at one of two places: creating (or revising) a model from data at hand, or drawing testable predictions from an existing model. We begin discussion at the latter point, assuming that a model of some natural phenomenon already exists.

Deduction and prediction
Deduction is a type of reasoning that in logic leads from a set of premises to a conclusion. In science, the logical premises constitute the model, which is a speculation of how things might work in nature, and the logical conclusion is a specific prediction that is testable by observation. One could also say that a deduction is the result of the

process of deducing, but that use would give two meanings to the same word, so we restrict "deduction" to the process itself.

Some writers encapsulate the deductive process as reasoning from generalities to particulars, but we think of it as a rearrangement of knowledge. In the terminology of logic, deductive reasoning extracts from a set of premises (the MODEL of Figure 1.1) one or more conclusions (the PREDICTION of Figure 1.1). No new knowledge appears in the prediction: everything in the prediction is already inherent in the model. The deductive process simply isolates part of the model or isolates several parts and then combines them. Consider a simple example:

> MODEL: the earth rotates on its axis, spinning counterclockwise when viewed from above the North Pole.

From this model, one can deduce the prediction that the sun (and the other stars, for that matter) should rise above the eastern horizon, travel across the sky, and set in the west. The model of course embeds some hidden assumptions, such as all these celestial bodies being fixed in space relative to the earth. Almost all models have implicit assumptions, and failing to recognize them could lead to problems in reasoning. The prediction deduced can be written down and checked empirically, which means that one uses the process of observation, including formal measurement, to see whether the predictions match reality as we view it.

Deduction is at heart a stipulated set of rules for assuring this relationship between model and prediction: *if the model is true, then the prediction deduced from it must also be true.* The rules take many forms, the oldest of which is the Aristotelian syllogism. In its commonest form, the syllogism produces a conclusion (prediction of scientific epistemology) from two premises (together making the model in epistemological terms). For example:

> PREMISE 1 (part of model):   If only Jack drives the van
> PREMISE 2 (part of model):   And if the van is on a field trip
> CONCLUSION (prediction):   Then Jack is on a field trip

Of course, science does not depend upon syllogistic reasoning, which is full of pitfalls because of the inherent imprecision of language. For example, consider this syllogism: unicorns all have horns (premise 1), and this animal has a horn (premise 2); therefore, this animal must be a unicorn (conclusion). The logical error in that syllogism is termed a fallacy of affirmation and is common in reasoning with language; just because A implies B (if it is a unicorn, it has horns) does not necessarily mean that B implies A (if it has horns, it is a unicorn). The error leads to apparent substantiation that unicorns exist. Anyone can see through the problem in a simple syllogism like this one, but much reasoning from scientific models is similarly linguistic in nature and so entails all the dangers of language itself.

Linguistic reasoning does not have to be in syllogistic form, as was demonstrated by Bertrand Russell. For example, it is possible to draw a conclusion from only one premise:

PREMISE: If a horse is an animal
CONCLUSION: Then a horse's head is an animal's head

Most scientific deduction is basically mathematical in nature, and, conversely, most mathematics taught in secondary schools is deductive in nature – Euclidian geometry and algebra, for example. Simple algebraic deduction can even be written in syllogistic form:

PREMISE 1:   If $x + y = 6$
PREMISE 2:   And if $y = 4$
CONCLUSION: Then $x = 2$

The deductive logic underlying the earlier example concerning the rotation of the earth on its axis was geometric.

The bridge between linguistic reasoning and mathematics is symbolic logic. Many such logic systems have been devised from different starting points, and in most cases different systems are easily shown to be equivalent. That is, given the same set of premises (together, the model of Figure 1.1), deduction by the rules of any given system leads validly to the same conclusion (the prediction of Figure

1.1). Boolean algebra is a form of symbolic logic that stands sort of midway between linguistic reasoning and traditional mathematics. The Boolean system uses the linguistic-like operators *and*, *or*, *not*, *if*, *then*, and *except* to relate variables such as propositions. Formal set theory is even more mathematical-like, using symbols for operators. All systems of deduction have in common the key property: if the premises (model) are true, then the conclusion (prediction) deduced from them is true, assuming that the deductive process scrupulously followed the rules of the system.

Observation and data

Observation in Figure 1.1 is the process that leads to data. "Observation" might seem a restrictive term, connoting merely noting visually what an animal is doing or some other aspect of the world. Nevertheless, we use "observation" as a general term to include all ways in which human senses are extended by instruments to record data. Ultimately, the investigator observes: for example, observes a dial or digital display on an instrument, or observes sound spectrograms made from tape recordings. Thus, observation in Figure 1.1 means any sensing and recording process that leads to data that can be written down or otherwise represented in hard copy.

The term "observation" might also be applied to the data produced by observing. In order to avoid confusion, we restrict our use of "observation" to the process and use "data" to describe the results of that process.

You might already have considered the question of why the arrow representing the observation process in Figure 1.1 points from PREDICTION, it being obvious why it points to DATA. The arrow originates at PREDICTION because the prediction specifies what kinds of data need to be observed. The prediction states what *must* be the case if the model is true, and the data show whether or not the prediction is upheld in the real world. Data that have nothing to do with the prediction might be an informative sidelight to a particular investigator's activities, but they have no direct bearing on the empirical test of the model. Nevertheless, observation with no particular model in mind can produce data that ultimately lead to a hypothesis. We have said that the investigator *usually* enters the epistemological cycle with a model to be tested or with data (often from the literature) for generating or revising hypotheses, so entering the scientific cycle through simple observation without starting predictions is an exception.

The word "experiment" does not appear in Figure 1.1. We have avoided that term because it tends to connote a carefully controlled laboratory environment in which every attempt is made to control extraneous variables that could influence the data. Some of biology – and especially the behavioral ecology practiced by the authors – involves mainly field studies. In some cases field studies also involve formal experiments, but in many cases they do not.

No fundamental difference exists, in terms of epistemology, between a laboratory experiment and field observations. Each approach to gathering data relevant to a prediction has its advantages and disadvantages. Laboratory experiments usually provide considerable control over extraneous variables that could influence the results, but the laboratory environment may produce artifacts. Field studies may be more natural and realistic, but generally they exercise little control over extraneous variables that could influence results. Both laboratory experiments and field observations play a role in research and can provide a particularly powerful approach when used in concert.

Comparison and decision

Sometimes, explanations of scientific epistemology fail to be explicit about the process of comparing the data observed with the prediction deduced from the model. This step is crucial to the workings of science, however, because it is not always evident whether the data are in agreement with a prediction. Extraneous factors *always* act upon any aspect of biology under observation, even in carefully controlled laboratory experiments.

As with deduction and observation, the term "comparison" could be applied to both the process and the results of that process. In order to avoid confusion, we use the word "comparison" to refer only to the process. The results of the process constitute the "decision" (regardless of whether the results fit the prediction).

The process of comparison commonly uses statistical methods for comparing data with predicted results. For example, a model might predict that older animals of some species tend to dominate younger animals. The data could show that this predicted relationship is an imperfect one, so the question becomes whether dominance structure is unrelated to age or is influenced by age as predicted or by some other age-associated traits. The investigator would probably employ some appropriate statistical test to see whether dominance relations were random or non-random with respect to age.

The comparison between data observed and the prediction deduced from the model yields a DECISION. Ideally, the decision is simply whether the data fit the prediction, but things do not always turn out so nicely. One may decide that it is impossible to tell whether a match exists. A common outcome of the process of comparison is that some expected difference is not statistically reliable, and yet a trend in the predicted direction is noticeable. Therefore, the difference could be real but not established by the data, either because extraneous variables unduly influenced the data (as commonly occurs in field studies) or because the sample size was insufficient to provide statistical reliability. In such cases, the main recourse is to gather better data, either with improved control over extraneous variables or with larger samples.

Yet another way in which comparison between predictions and data can fail to yield an unambiguous decision about the model is when some predictions are supported and others are not. Any validly deduced prediction that is rejected by empirical data falsifies the model, but when the deductive chain is not tight and other predictions are consistent with data, researchers sometimes refer to "partial confirmation" of the model. This situation usually suggests that the model is not quite right but has merit that could be improved by modifications based on careful scrutiny, without having to "go back to the drawing board."

Induction and model

The decision resulting from comparison between data observed and the prediction deduced from the model dictate the next step in doing science. As noted earlier, three kinds of decisions are possible: (1) the data do not resolve the question because they are inadequate or conflicting, (2) the data do not match the prediction, or (3) the data confirm the prediction. In the first case, if the data are not sufficient, then nothing can be done except to go back and gather data adequate to the task. On the other hand, if the results testing different predictions conflict, then the model is probably not quite right and needs to be modified.

Suppose the data show unambiguously that the prediction cannot be correct: the data are simply not as predicted if the model is true. Only one explanation of this situation exists: the model is false. We say in science that the data reject the model (because they do not match the prediction deduced from the model). In this case, if one is to explain the phenomenon under investigation, it is necessary to produce a new model, or at least revise the old one, and then proceed with making and testing new predictions.

The creative process of proposing how nature works involves induction. Induction is sometimes characterized as reasoning from particulars to generalities, but that catchphrase seems vague and in any case may not always apply. We prefer to think of induction as a cluster of very complicated creative processes in which the thinker identifies possible patterns from available facts and proposes causal relationships that might explain the patterns.

No rules for creative induction exist, and the many books written on the subject seem to agree that induction is not one but many complex mental processes that find a pattern where none was recognized previously. Most of the famous models of science have come from people who reflected on disparate empirical data and somehow united them into a coherent framework. For example, Charles Darwin realized that the traits of parents and their offspring tend to be similar (genetic inheritance was not yet understood), that more offspring are born than survive to reproduce themselves, and that survival probably depends at least partly on the traits of the individual. From these empirically verifiable facts, Darwin reasoned correctly that if the survival traits are heritable, then evolution must occur: his model of natural selection. To recount another famous example, Danish physicist Niels Bohr mused over the emission spectra of elements. When one burns a substance, it emits light of specific wavelengths, the combination of wavelengths being unique to every different element. From a vast storehouse of emission spectra accumulated by empirical scientists, Bohr conceived of his atomic model of a positively charged nucleus surrounded by negatively charged electrons of different energy levels.

The examples from Darwin and Bohr are what historian of science Thomas Kuhn has called "paradigm shifts" or "revolutions": whole reorganizations of thinking in a given area of science. Kuhn first believed that progress in scientific understanding was completely dependent upon such revolutions but later came to realize that stepwise revision of models also moved science forward (see Kuhn 1996). An entire spectrum exists from minor honing of models through substantial revisions and generalizations to major shifts in paradigms, and all have their place in the progress of science. Few practicing scientists will bring forth revolutionary new ways of viewing some natural phenomenon, but each scientist should strive to keep an open mind and induce new, viable ways of uniting disparate facts through induction.

Uniqueness of models

Because induction is a type of creativity, scientific models that result from it are unique to their creators. This assertion is controversial, but most apparent exceptions to the asserted uniqueness turn out not to be exceptions at all. The whole issue is somewhat tangential to the main point of devising or revising models, but the notion of uniqueness is important in emphasizing the creative process of induction. Therefore, we pause to consider several apparent exceptions that help to illuminate the nature of both the inductive process and the models that result from it.

First, radar was invented nearly simultaneously and independently on both sides of the Atlantic at about the time of the Second World War. Radar was a marvelous technological innovation, but it was not a scientific model. Modern inventions are hardly the product of tinkering, and they usually incorporate all sorts of science in their development. Inventions are not scientific models, however, and essentially the same invention may be developed independently by two different parties.

Second is the famous case of independent creation of calculus by Leibnitz and Newton. Indeed, these two geniuses quarreled over priority, each believing that he had conceived the concept of calculus first. Mathematical techniques are not, however, scientific models. They cannot be rejected by empirical facts, but rather they are logic systems that must be evaluated by their internal consistency.

Third, the planet Neptune was discovered independently and nearly simultaneously by different astronomers. This was a scientific discovery, but it was not the creation of a causal model of how nature works. The discovery of Neptune was, in fact, a datum in the sense of Figure 1.1 and had been predicted on the basis of a model erected to account for perturbations in the orbit of Uranus. Two scientists can obviously make the same empirical discovery wholly independently of each another.

Last, the most celebrated proposal for independent creations of the same model is the idea of natural selection proposed simultaneously in 1858 by Alfred Russel Wallace and Charles Darwin in separate essays published together. Darwin himself was convinced that the two men had said exactly the same thing and had even used some of the same examples, such as domestication as a form of selection. Wallace and Darwin certainly had similar ideas and used related examples, but a close comparison of their original essays shows that the models were distinctly different. Wallace conceived of selection as acting mainly on entire species or groups of animals, whereas Darwin initially focused entirely upon selection among individuals of the same species. They even used the case of domestication differently in their conceptions. Wallace viewed the reversion of feral animals to the wild form as an example of selection, whereas Darwin viewed the original domestication of wild animals as a type of selection imposed by humans – in effect, an analogy with natural selection. The general ideas of selection were similar but not really identical.

In sum, two scientific models may appear similar and even be virtually identical in some respects. Nevertheless, it is unlikely that all underlying premises will be the same, so each model is truly unique.

Some consequences of the uniqueness

A point made by the examples in the foregoing section is that many different kinds of creative products and discoveries become part of the corpus of science. Most of these creative contributions, however, can be (and sometimes are) duplicated independently by two or more people, whereas the creation of a causal model is unique to the individual. If Bohr had not proposed his model of the atom, then someone else probably would have had a related conception sooner or later. The other model would not have been identical to his, however, so details of the subsequent development of chemistry in the twentieth century would have been at least a little different from the history we view today in retrospect.

Spurred by the feminist movement, especially in the USA, a debate has arisen as to whether such a thing as "feminist science" exists and, if so, how it differs from "other" science. Recognizing the individual uniqueness of models helps to clarify one issue of this debate, for it points to the question of whether creative processes of women and men tend to differ in consistent ways. If so, then it might be possible to separate "feminine" and "masculine" versions of science based on the characteristics of the models offered to explain the outside world. It will likely be difficult to establish whether women and men differ in some way in their scientific creativity, however. The differences among individuals may prove much larger than any difference between the sexes. Furthermore, it will not be easy to dissociate differences in creative processes from differences in what females and males find to be interesting or informative topics at any given time. Whether differences in scientific creativity occur is basically a question left to sociologists, historians of science, and philosophers of science.

The uniqueness of models – especially those rare models that are both generally applicable and widely verifiable – bestows high praise upon their creators. It is not surprising that journalists, philosophers, psychologists, and colleagues have tried to find a key to the inductive powers of great model makers. This kind of creativity, like other kinds, remains somewhat mysterious, however. Great thinkers in science usually employ wonderful mixtures of mental processes: recall of disparate facts, extrapolation beyond the established bounds of some relationship, notice of the trivia overlooked by others, perception of similar patterns in apparently unrelated phenomena, identification of connecting points between things that seemed to have no connection, and so on. No guide to induction exists. It has no rules like those of deductive logic systems.

A well-conceived model rarely fails entirely to predict reality. More commonly, empirical data show a model to be wrong, but not dramatically wrong. In these cases, a wholly new model may not be necessary to account for a natural phenomenon; a revision of the original model may be what is needed. Inductive processes are again called upon for the revision, but the task is usually not as difficult as devising an entirely new model.

Models: hypothesis, theory, law

We have left the discussion of models purposely incomplete up to this juncture so that a separate section could be devoted to a critical issue: what happens when the data observed do match the prediction deduced from the model? The first temptation might be to say that the empirical test has proven the model to be true, but this cannot be the case for a simple reason: false models can make verifiably true predictions.

Consider the model that the earth is fixed in space and the heavens rotate about it from east to west. This model predicts that the sun should rise in the east, move across the sky, and set in the west. As early risers know, that prediction can be verified empirically on any clear day, and yet science holds that the model from which it was validly deduced is false. The reason we reject that model is that it makes certain other, more subtle and detailed predictions that are not upheld by comparisons with other data.

An inevitable consequence of scientific epistemology is that no model can ever be proven true. As an anonymous wag put it, "The answer is maybe, and that is final." This inevitability is inherent in the method of science and no one has ever devised a way around it. The only certainty achievable is rejection of a false model, as true models generate only true predictions. False models can generate predictions that fail to match reality, and so may be discovered to be false, but false models can also generate true predictions. Therefore, finding that observed data match a deduced prediction merely suggests that the model might be true. Data can confirm a prediction, but they cannot prove as true the model from which the prediction was generated.

The diagram of the epistemological cycle of science (Figure 1.1) might be a little misleading when it comes to a decision that the data match the prediction. Here, no new or revised model is called for, but merely further testing of the model that survived the current test. The arrow from DECISION to MODEL is correct, but it indicates that the process of science cycles anew. In this case, the label "induction" may not be applicable, and the only process represented by the arrow is that of beginning the cycle of epistemology for another round. Nevertheless, it has been argued that because few models correspond exactly with reality, practicing scientists really do modify their underlying models at least a little with every pass through the epistemological cycle. If that is an accurate view of how science really works in practice, then "induction" may be an appropriate label for the arrow in Figure 1.1 almost all of the time.

Each pass of the cycle that fails to reject the model under scrutiny increases our confidence that the model might be true. Labels are commonly used to indicate the degree of confidence placed in a model. No special name exists for very preliminary models, especially those that have not even been checked against already available facts. We refer to such models informally as ideas, suggestions, hunches, possibilities, and so on. "Hypothesis" is the name given to a well-stated model that is already known to be in accordance with the existing body of evidence and may even have survived some formal empirical tests. "Theory" designates a well-tested model in which the scientific community places a great deal of trust because it has survived repeated attempts to prove it false. Finally, "law" applies to only the most thoroughly tested models that are so robust that they are unlikely ever to be rejected entirely. Occasionally, laws fail to make true predictions in unexpected realms, so they have less than universal generality. In such cases, the laws might turn out to be special cases of a more general model – as classical Newtonian mechanics proved to be a special case of Einstein's general relativity model.

Roles of the individual investigator

Every scientist does at least some of the things shown in Figure 1.1 – deduction, observation, comparison, induction – but not necessarily all of them, especially on a given research project. Physicists, for example, commonly characterize themselves as either theoreticians or empiricists, the former spending most of their energies devising mathematical models and the latter devoting themselves to experimental gathering of data to test the models. Similar divisions, although usually not so clear-cut, occur in biological disciplines such as community and population ecology. Even if such a division is not general for all research endeavors, then biologists and other scientists may restrict their activities to one or several parts of the epistemological cycle for a given research project.

Furthermore, most biological research projects are multifaceted, involving several related but different phenomena to which several different models may apply. This multifaceted approach is especially common in field studies, as of behavioral ecology, for example. This means that the investigator may be involved with different parts of the epistemological cycle for different aspects of the study, as in gathering data about one aspect while attempting to devise a model about another aspect.

Is natural history really science?

Some people would denigrate natural history as unscientific, and although some truth lies in that accusation, the biologist needs to avoid the danger of throwing out the baby with the bath water. The viewpoint of scientific epistemology put forth earlier is known by philosophers as the hypothetico-deductive method and is characterized by others as "strong inference" (see Platt 1964).

Until the past few decades, the venerable study of natural history operated largely on what has sometimes been termed the "advocacy" procedure, or "weak inference." A researcher gets an idea from observations of nature and proposes it publicly, whereupon a peer challenges the idea and the researcher replies. The obvious danger of advocacy science is that the discipline may become divided into camps of belief, each becoming more caught up in defending its model than seeking the underlying truth. Nevertheless, advocacy continues to play a large and important role in science because it has the psychological effect of impelling researchers to scrutinize one another's evidence with a zeal that may be lacking in assessing their own data. As in law – where advocacy is the very basis for uncovering the closest approach to truth – vigorous defense of one's pet hypothesis can have a salutary effect in science if the defenders are willing to admit when they have been proven wrong.

We believe that the contrast between strong inference and advocacy has often been drawn too sharply. Indeed, an irony is that many of those pushing the former to exclusion of the latter have actually been using the advocacy method to promote strong inference. We view the issue mainly as one relating to the maturity of a question or phenomenon. Initial models induced from natural history observations are new ideas that are valuable to put before peers with an explication of the data that led to them. It is often difficult in the early phases of a given line of investigation to think of alternative models and devise critical tests to separate them. Advocacy of a new model stimulates other researchers drawing on different experiences to propose and defend alternative possibilities. When this point has been reached, the discipline should then begin to move on to strong inference – by carefully specifying the core differences among alternative models and deducing predictions that could lead to rejections, hence narrowing the field of possibilities. A line of research reaches maturity when everyone involved is trying to test models rather than advocating them.

PLANNING RESEARCH

In parody of a line from Scottish poet Robert Burns, one can say that the best-laid schemes of mice – and the biologists who study them – gang aft a-gley (often go amiss). Nevertheless, planning is the essence of success in science, because when things do go in an unanticipated direction, it is easier to make viable adjustments in a well-conceived plan than when the approach was hazy to begin with. It is especially important for the field biologist to plan carefully, as there may be no second opportunity, particularly if a foreign expedition is involved.

Simple good fortune can play an important role in science, but the general truth that chance favors the prepared mind cannot be overlooked. The prepared mind in science ranges broadly in background reading and familiarity with other disciplines, a point reiterated occasionally in the following sections. Graduate students frequently make the mistake of concentrating too narrowly on the material of their chosen discipline. In fact, we recommend strongly that they seek out interdisciplinary seminars, go to talks outside their immediate fields, and read journals and books in other disciplines. Primatologists should know something about birds, and ornithologists something about primates – and insects, and amphibians, and many other animals. Physiologists need to understand the design of behavior studies, and behaviorists should have some knowledge of physiology – and ecology, and genetics, and many other areas. A general background in the sciences and social sciences pays good dividends when trying to find a problem, to formulate a model, and to devise testable predictions in your own area of science.

No one can effectively tell another what to study or details of how to do it, but in the next sections we make some suggestions that researchers may find helpful. We think that some aspects of doing science are noticeably more difficult than others, and we emphasize the more difficult aspects.

Finding a problem

Graduate students in particular often experience difficulty in getting started on their dissertation research because they do not yet know how to find a problem worthy of investigation. They may be familiar with many theoretical issues in their field but do not yet know how to translate the generalities into specific problems for investigation. Or, they may have an intense interest in a species or group (e.g. social insects, marine mammals), or in a general subject area (e.g. mating systems, foraging), but cannot identify a specific problem with theoretical implications. It is important to get over such a block as soon as practicable so one can begin planning a specific research problem. Here, we offer some comments that may help to get past that block and identify a feasible problem.

First, a graduate student should go to his or her thesis advisor – that is what advisors are for. Do not make the mistake of believing that a beginning graduate student must have a well-articulated research problem ready for scrutiny. Instead, be prepared to explain those things that seem to motivate you most: field versus laboratory work, observation versus experimentation, the kinds of animal or preparation that you find interesting, and the sorts of phenomenon that grab your attention. In other words, try to circumscribe the sphere of possibilities within which you might work. Good advisors want not to assign specific problems to their students but rather to develop problems in concert with and tailored to the individual advisee. A first experience in science, such as a senior thesis, usually requires the advisor to impose a lot of structure based on his or her own research experience. As you progress to master's and then doctoral work, your advisor will expect you to show greater independence and to arrive at the office door with more specific problems in mind for discussion.

The general strategy of successful scientists is to choose the most important problem that they think they may be able to solve. At the extremes of the spectrum of potential problems are those that are too large or too difficult for a single research project and those that are so safe that they are virtually trivial. For example, the problem of how and why sexual reproduction evolved is not the stuff of which a good graduate thesis is liable to emerge, although a more senior investigator might wish to build an ongoing research program around such a problem. Conversely, one would not like to choose some small, specific problem such as how fast a given species of bird can fly, unless that question contributed clearly to some larger theoretical issue.

Many ways exist to find a problem worthy of investigation. Sometimes a problem is more or less provided to an investigator, especially the thesis problem of a senior undergraduate or beginning graduate student. In other cases, particularly involving technological or other practical research, a funding source may encourage selection of one of the problems that the agency or foundation would like to have solved. Although having a funding source dictate a problem may sound, and indeed can be, crass, it is not necessarily so. For example, one of us accepted a contract from the US Fish and Wildlife Service to investigate the cause for and possible abatement of the frequently fatal attraction of endangered seabirds to manmade lights on the island of Kauai in the Hawaiian archipelago. The contract negotiated included a substantial component of basic research in avian vision, as well as application of the findings to solve the practical problem.

More commonly, problems for investigation arise from ideas, encouragements, or interactions with other scientists. For example, many advisors may present their graduate students with a list of possible problems or a problem area within which specific problems might be identified. Another source of ideas are the gurus of a discipline: the senior scientists who write the review papers and books that explore some area of science and suggest unsolved problems within that area. A third, and perhaps the richest, source of ideas for problems comes from the discipline in general: read the technical journals and email communications; go to professional meetings where you can hear oral papers, read poster papers, and talk with other practitioners; listen to talks on campus; and, finally, discuss issues with peers and others on the home campus.

No one is uninfluenced by what has transpired in a given field and what is going on currently, so there really is no such thing as finding a research problem *de novo*. Nevertheless, a lot can be said for Louis Agassiz's admonition to "learn from nature, not books." Many of the most creative biologists, especially in behavioral ecology and other disciplines with strong ties to the natural environment, come to their problems by firsthand experience. They observe something in the field and ask themselves questions such as "How does that work?" or "Why does an animal do that?" The great ecologist Robert Macarthur once asked a question that had not quite occurred to anyone previously in just this form: "Why don't predators overeat their prey?"

We do not recommend that a graduate student should attempt to set the world afire by choosing a problem for thesis research that everyone immediately recognizes as of critical importance to the theoretical corpus of science. It should become the mindset of every scientist, however, to find as good a problem as he or she thinks *solvable*. It has often been asserted that the most creative aspect of science lies in asking a good question and identifying a good problem. Try to launch your career by asking good questions of nature right from the very beginning.

Try also to calibrate your own expectations and perceptions of a good research project with those of your advisor, committee members, and other more experienced colleagues. An interesting research question might be solvable once a particular technique is mastered, validated, and applied. Nevertheless, if it is going to take four or five years to develop the technique, then it may not be an appropriate problem for a graduate student to tackle. Frequent consultations with your advisor and other faculty establish the good communication skills that are necessary to the enterprise of science.

Formulating a model

The second major hurdle in planning research is to formulate potential explanations for the phenomenon chosen for study. It is usually at least as difficult to build a testable model as it is to identify a good problem, but sometimes the problems themselves suggest potential solutions. Indeed, many biologists report that the problems they tackle remain subliminal until they have articulated a possible solution. Scientists think in terms of cause and effect, and so once they propose to themselves a causal relationship, the problem they were subconsciously wrestling with becomes explicit.

Many paths exist for developing a model for one's chosen problem, and these paths range from adopting a preexisting model (say, from the literature) to constructing one wholly from scratch. In most cases, some complicated path between these extremes is followed, amalgamating elements and aspects for various sources. One ordinarily adopts a model *in toto* from the literature for the explicit purpose of testing an idea that has interested researchers in the field but is neither well confirmed nor convincingly rejected. When the problem to be tackled is already well recognized but available models have proven inadequate, a new model is frequently developed by modification of a preexisting one.

In cases where a familiar problem has been broadened for revived study, preexisting models may also be useful. One may be able to extrapolate from older, more limited models. Consider the following example. It is well known that in most species of hawk, the female is larger than the male. A standard model attempting to explain this sexual dimorphism is that breeding pairs cannot range far enough from the nest to cover a sufficient area to sustain both mates if they hunt the same prey. As the size of a predator is related positively to the size of its prey, other things being equal, the mates can lessen competition for food by hunting different prey. They become adapted to this ecological divergence by also diverging in size. This model does not explain why the female is larger than the male, but that issue is not critical to the example. If one broadens the problem from sexual dimorphism of size in hawks to birds in general, then extrapolation of the model to explain the case of hawks might be a useful start for dealing with the more general problem. One could propose, for example, that in any avian species where males and females differ in size, the difference is traceable to a need for mates to minimize competition for some resource, which is not necessarily food, as it is in the case of hawks.

When the problem to be solved is not a familiar one in the literature, preexisting models may still prove useful in developing a new model by analogy. Borrowing an idea from one field of study to use in another is actually quite common in science, especially where mathematical models are concerned. Sometimes it is just the heart of a quantitative relationship that is borrowed and adapted to a new situation, but it becomes the basis for a causal model in the new setting. To give a real example, Ivan Chase is a sociologist who studies animal behavior. He was familiar with a sociological model called the vacancy chain, in which when the highest executive of a company retires, dies, or moves to a new job, the second in command succeeds to the top post, the third member on the ladder assumes the job vacated by the second, and so on down the chain. One strategy for getting ahead in business is to work in a subordinate position when an expectation of moving up the vacancy chain at some future time exists. This strategy may be more productive than searching externally for a different position of higher status. Chase recognized that aspects of the vacancy chain model of sociology might be applicable to the behavior of hermit crabs. Instead of wandering aimlessly to find a larger empty mollusc shell to occupy as the crab grows, it might prove viable to stay near slightly larger crabs. The smaller crab can appropriate the vacated shell when the larger crab leaves to find a yet larger shell.

It is not always possible to modify, extrapolate from, or make an analogy with a preexisting model, so one must nurture other strategies in formulating a new model. One strategy is to discuss the problem to be solved with peers and more experienced scientists. Sometimes other people will have suggestions that can, with development, become the basis for a new model. Perhaps more frequently, the questions of others can potentiate one's own thinking and elicit new ideas about possible explanations for the phenomenon to be studied. Group discussions can often be particularly useful in this regard.

Another strategy for formulating a new model is to generalize from a particular instance, say from something observed in nature. Niko Tinbergen's discovery of "cutoff appeasement" signals is an example of such generalization. While reviewing his films of mated black-headed gulls, Tinbergen noticed that in propinquity, the mates suddenly turned their heads away from one another, an act he later termed "head-flagging." The turning served to hide the threatening black facial mask and the main weapon (the bill) of the bird and cut off possible aggression from the mate. Tinbergen later found this same principle at work in many species, where an individual would suddenly hide a weapon or aggressive signal in situations where it was useful to defuse possible aggression from another animal. One instance, from a piece of movie film, became a general principle of ethology.

We offer a suggestion to those developing models: think broadly, formulate general classes of possible explanations, and whenever possible, develop two or more different models to account for the same phenomenon. Perhaps no stronger approach to science exists than the pitting of alternative models against one another. Here is a published example of multiple-model formulation. One of us, who had observed blue jays vocally mimicking the calls of several species of raptorial birds, wondered why the jays did this. Instead of trying to think of ad hoc explanations, the investigator in this case asked first what the underlying question should be in order to develop classes of models. Once considered in this way, the problem in general form became: "Who is the intended listener of the jays' mimicked calls?" Three types of potential listener were identified, with a fourth class for any sort of model not covered by the first three, so four classes of models in all were articulated. In brief, the first is that conspecific companions might be the intended listeners (one specific model being that a jay identified to companions a predator it saw by mimicking a call of that predator). The second class of models listed the predator itself as the intended listener (one specific model being that a jay could distract a predator from hunting by mimicking its own call). The third class proposed that a third species was the intended listener (one specific model being that the jay could frighten off competitors by deceiving them into believing a predator was present, thus allowing the jay access to some resource being monopolized by the competing species). It is always useful to have a "wastebasket" category, so in this example a fourth class of models was that there was no specific intended listener of mimicked calls per se; a specific model might be that jays imitate a variety of sounds as one mechanism of achieving vocal variety, and calls of hawks just happen to be easy to mimic for some reason. In case you are wondering, no resolution of the problem has appeared as yet, although in response to reading the classes of hypotheses, a field observer reported seeing jays give hawk imitations that frightened grackles off a picnic table, the jay going to the table and feeding as soon as the grackles fled.

Devising testable predictions

The last difficult aspect of planning research is to devise testable predictions from the model or models to be evaluated empirically. In broad sweep, two basic tactics that biologists in general and especially behavioral ecologists use in testing models are formal experiment and observation. As Niko Tinbergen pointed out, observation is a powerful method of field biology because nature eventually performs experiments for the patient observer. The important aspect of a prediction is that it must be capable of falsification by empirical data.

As with finding a problem and formulating models that might explain the chosen phenomenon, devising predictions can be aided by existing literature. The most straightforward way of devising a prediction is to adopt a standard testing method that has been used by many others to test similar models. To take a simple example, if the model states that a particular vocalization acts as an alarm signal, making companions flee for cover, then this model predicts that playing back the call through a hidden loudspeaker will elicit fleeing.

Similarly, even if no standard methodology exists, it may be possible to adopt ideas in the literature for applying to your particular model. For example, suppose that your model states that younger monkeys are especially exploratory of new potential foods. One might begin to test this model through observation by recording for animals of different ages what kinds of plants or animals they attempt to eat, predicting that the diversity of items will be larger in the younger monkeys. Even if the data are consistent with the prediction, they do not account for why younger monkeys have more diverse diets than older monkeys. Therefore, one might also devise an experiment based on previous work with Japanese macaques, where primatologists provisioned troops with unfamiliar foods. Here, one would predict from the model that the younger animals are most likely to be the first to try the new food provided.

In many cases, past literature will be of limited help, so one must devise testable predictions more or less *de novo*. This task is usually not as easy as it might seem at first blush. One approach is to scrutinize the model and follow any line of deductive reasoning that might lead to an observable result. For example, if the model states that butterflies congregate at wet spots on the ground in order to obtain one or more critical nutrients, then a prediction is that all such spots will contain at least one such nutrient – a prediction testable by chemical analyses of samples from the spots.

Another approach to devising testable predictions is to make further observations while keeping in mind the model to be tested. New observations may suggest predictions that had not been evident through initial scrutiny of the model. Continuing the example of butterflies, further observations in the environment might turn up wet spots at which butterflies do *not* gather. The model predicts that these spots should lack the nutrients found in spots where the insects do gather. In this case, one could consider this a new prediction or, taken together with the first one, a stronger single prediction from the model.

Investigators should always consider the possibility of testing models, especially functional and phylogenetic models, using the comparative method. For example, one of us studying vocal communication in the Mexican chickadee could not help but notice the buzz-like quality of the calls and wondered whether this phenomenon was an adaptation to the montane coniferous forests to which the species is confined. This simple model predicted that congeners restricted to coniferous habitat would also have buzz-like voices, whereas those of broad-leaved forests would not. The results of testing the prediction by looking at available information in the literature point up the ever-present problem with comparative studies. The predicted correlation was substantiated; however, it turned out that those species known to have buzz-like voices not only were restricted to coniferous forests but also were related more closely to one another than they were to any species whose voices were not buzz-like. In such cases, it is necessary to look at a wider comparative sample to dissociate similarities that might be adaptations from those that are due solely to common ancestry (see *Comparative studies*).

Whatever the strategy used for devising predictions from the model to be tested, it is always good science to derive multiple predictions from the same model. Although results consistent with a model increase our confidence in that model, they cannot prove it to be true; therefore, the more predictions that are tested, the more confidence we can place in the model. For example, the model that blue jays mimic the calls of predators in order to warn conspecifics of the presence and type of predator makes several testable predictions. The jays should mimic only predators that prey upon blue jays (including their eggs or young). Also, when a mimicked call is given, that species of predator should be present. Furthermore, mimicked calls should be given only when conspecific companions are present to be warned. Inherent difficulties exist in testing all such predictions convincingly, but if several predictions are derived from the same model, then the chances of obtaining a good test of a model are obviously enhanced.

Finally, the previous section emphasized the importance of formulating multiple models to account for a given natural phenomenon, so it is a useful strategy to consider together two or more models when devising predictions. A powerful way to proceed in science is to eliminate models by finding situations in which two or more models make conflicting predictions. This situation is often termed the "critical test." For example, it was found that individual black-capped chickadees that were high-ranking in dominance sometimes tarried at a feeding station to open sunflower seeds, whereas lower-ranking individuals never did this. As with all correlations, cause and effect between the correlated variables (seed-opening and dominance in this case) are uncertain. At least two models are possible: high-ranking birds can afford to tarry on the feeder without fear of being supplanted by subordinates, and high-ranking birds tarry as a way of reinforcing their dominance by requiring subordinate companions to wait longer before visiting the feeder. These two models make opposite predictions about what birds will do on the rare occasions that they visit the feeder alone without the other members of the flock being present. If tarrying is a result of dominance, then any bird should at least occasionally tarry when alone because no others are present to supplant it. Conversely, if tarrying functions to prevent others from immediate access to the feeder, then lone birds should never tarry because no one is present whose access can be denied. It is not always easy to find a critical test between two models, but it is always worthwhile to strive for one.

Comparative studies

Studies comparing different species are special, often entailing considerations in addition to those found in straightforward observational and experimental studies. Various reasons exist for studying a variety of species. A common one is to find a species best suited for investigating a particular problem, the so-called August Krogh principle (after the Danish Nobel laureate in physiology). Many disciplines within biology concentrate on a particular species or preparation as a model for more general phenomena: *Drosophila* in genetics, the squid giant axon in neurophysiology, and the laboratory mouse in immunology are familiar examples. More often, however, comparative studies are undertaken to tease apart the influences of evolutionary relationship from those of adaptive selection upon some trait of interest.

The confounding of relationship and adaptation occurs when the ancestor of a group of related species evolved an adaptation that is perpetuated in the descendent lines even if no longer required by some of the derived species. For example, all pigeons and doves drink using a method unusual among birds: inserting the bill into water and pumping it up to the esophagus. This behavior allows water to be taken more quickly than the method used by most birds of scooping water into the lower mandible and raising the head to swallow. The pigeon method is believed to be an adaptation to desert conditions, as the birds must drink quickly at a water hole where predators lurk. Yet even species of pigeons and doves that live in well-watered environments use this method, presumably inherited from a desert ancestor that gave rise to the entire family of species we know today. Comparative study can help us to evaluate the hypothesis that the unusual drinking method is an adaptation to arid conditions by looking more widely at drinking behavior of unrelated species that live in similar environments. In this example, it was found that certain finches also drink by the pumping method, namely those species living in arid conditions, whereas their close relatives living in other habitats drink by the more usual avian method.

Put simply, the comparative method seeks to establish whether unrelated species living in similar environments possess similar traits. If their close relatives living in different conditions have alternative traits, then a good case for adaptation is secured, although it may happen (as in the drinking of pigeons) that closely related species all have similar traits because of common evolutionary ancestry. It follows logically that the minimum design for a comparative study needs to compare at least two closely related species living under different critical conditions from each of two relatively unrelated taxonomic groups. Only with this minimum design can one confirm or reject the hypothesis of a trait–environment correlation that would indicate adaptation.

Interrelationship between field and laboratory

Some things have already been said about field and laboratory research, but more needs to be added. The first part of this chapter pointed out that in terms of testing hypotheses, no fundamental epistemological difference exists between field and laboratory research. We also said that straight observation, which usually comes from the field, is a viable route to generating new models. Such straightforward observation tends to occur before a biological phenomenon enters the ''strong inference'' stage of testing, which is often more oriented towards the laboratory. We have even mentioned that field experiments and laboratory descriptive observations often have their place. What remains is to emphasize that the combination of field and laboratory study can be an extremely important component of planning research.

Consider a specific example. In captive colonies of callitrichid monkeys of several species, it was found that only one female usually bred. Hormonal assays of other females housed with the reproductive female revealed that their reproduction was being suppressed (behaviorally or via pheromones). When the females were separated, non-cycling females began to cycle almost immediately. These laboratory studies are reliable, but drawing the generalization that such reproductive suppression is an important factor in sociosexual organization of the monkeys under natural conditions would be unjustified. In fact, later studies in the field found the presence of more than one breeding female in many callitrichid monkey groups. Without the combination of laboratory and field studies on the same topic, our understanding of these matters would be quite incomplete.

Whereas most laboratory results need to be checked against naturally occurring situations – this being as true in medicine as in ethology – the mandate goes as strongly in the other direction as well. Field studies are usually more difficult to control than laboratory experiments, so phenomena must often be studied ultimately in the laboratory in order to obtain definitive tests of predictions. Furthermore, almost all modern field study involves a laboratory component in the spectrographic analysis of acoustical tapes, computer analysis of quantitative data, and so on. Indeed, students of avian behavior often find that things happen too fast for reliable observation; field work becomes relegated to making videotapes, and the actual observation of behavior occurs later in front of a monitor.

In short, an increasing (and we think healthy) interplay exists between field and laboratory activities for a given project. Your research planning can be enhanced by paying equal attention to what can be accomplished in the laboratory and the field, and how those accomplishments interrelate and mutually inform one another.

Other aspects of planning research

Some other aspects of doing science are also important to planning research. The principal endeavors of planning – which are also the most difficult parts of doing science – entail finding a problem or phenomenon to be explained, formulating models that might explain it, and devising testable predictions from those models. These principal aspects of planning merely begin the cycle of investigation (Figure 1.1); one must proceed to make the observations that yield data, to compare the data with predictions in order to decide whether the model has survived the test or is rejected by it, and then to make further predictions from surviving models or to revise or replace models that have been rejected. These other aspects benefit from advanced planning.

It is one thing to have a prediction that is testable in principle, but one needs to ensure that a practical way exists by which to collect the relevant data. It is obviously fruitless to predict some difference between male and female if the sexes cannot be distinguished. When Stephen T. Emlen tested the model that indigo buntings, nocturnally migrating birds, used stellar patterns in the vicinity of the polestar to orient, he made a straightforward and testable prediction: the birds should orient correctly if the only cues available were the star patterns. To test this prediction, however, he had to find a way of presenting the night sky when other cues such as prevailing wind patterns were not available; this he did by caging birds in a planetarium. This general approach by itself was not sufficient; Emlen also had to devise a way of blanking out portions of the simulated sky and to record the direction of movements of the migratory restlessness of the birds. It turned out that some planetarium projectors can show only restricted portions of the sky, but to record direction of the birds' movements Emlen had to invent a special recording cage. All such details need to be planned as much in advance as feasible and even tested by pilot experiments to ensure that a study will run well.

Especially in studies of behavior, the observer needs to be aware of the possibility of his or her own influence on the animals studied and the possibility of bias in judgments required in recording data. Our colleague Charles T. Snowdon related this example of the first problem: "In a study I did in 1968 I found that the dominance rank of chickadees (based on supplanting) predicted feeding order and duration at feeders far from me as an observer, but that the reverse hierarchy was found at a feeder close to my position." With regard to the second problem, if judgment is called for in recording data, then explicit tests should be devised for interobserver reliability.

Analysis of data might seem a topic to be deferred until the results of the observing process are complete, but collecting data without a definite analytical procedure in mind is a grave mistake. Of course, one needs a formal protocol for collecting data to ensure that a standard procedure is always followed. Indeed, it is not only formal experiments that require a protocol but also systematic field observations, where the complexity of the endeavor makes a protocol even more important. Beyond having a protocol, however, one needs to plan in advance how the data will be analyzed. This subject is often considered in statistics courses under the rubric of "experimental design," but it is just as important (or more so) to design the collecting of observational data in the field so that they are analyzable with valid statistical methodology. The nuts and bolts of how to plan data collection for ultimate analysis is a topic that fills entire volumes and cannot even be summarized here. We can point out, however, that failure to plan data analysis is often the Achilles' heel of an otherwise well-conceived research project.

Implementing the plan

Subsequent chapters of this book offer aids to implementing planned research and related subjects. Every scientist must engage in a lot of writing, from statements of models, predictions, and protocols to composing a curriculum vitae. An appendix is therefore devoted to hints on writing clearly. Chapter 2 explains how to prepare a proposal for funding the planned research. After the research is completed, one must write up the results (as a dissertation or for publication), the subject of Chapter 3. Presenting research results as an oral or poster paper at a professional meeting, or as an hour-long seminar for departmental colloquia or "job talks," is covered in Chapter 4. Finally, the last chapter deals with preparing a curriculum vitae when applying for research positions and grants.

Finally, Appendix B on ethics considerations is a completely new addition to this edition. We mention ethics as appropriate in each chapter, but the topic is worthy of a separate discussion.

2 How to write a research proposal

Format

Criteria for evaluation
    What is the intellectual merit of the proposed activity?
    What are the broader impacts of the proposed activity?

Feedback

Proposal content
    Significance of a title
    Identifying objectives
    Integrating your research with existing knowledge
    Hypotheses and predictions
    Methods
    Significance
    Literature cited
    Summary
    Budget and budget justification
    Other funding
    Appendixes
    Table of contents

The submission process

The review process

What to do while waiting for a decision

Benefits which cannot be repaid and obligations which cannot be discharged are not commonly found to increase affection.

Tacitus (*c.* 55–120)

Research proposals are of course required by funding agencies, but graduate students usually must submit a formal proposal of thesis research, and there are other occasions upon which established scientists must write proposals. This chapter explains how to prepare and submit a research proposal, using the Doctoral Dissertation Improvement Grant format of the National Science Foundation (NSF) as a specific example. Discussed are the criteria used to evaluate a proposal, the usefulness of testing your plans on others before submitting, the importance of pilot studies and preliminary data, details of the content of a proposal, and various tips and suggestions. The heart of any kind of proposal is tripartite:

TIP

Explain what you want to do, how you will do it, and why it is important.

It is usually assumed that a disorganized or poorly written proposal reflects a disorganized or poorly conceptualized study. Therefore, organization, content, and clear writing are essential to the success of a proposal, whatever its purpose. This chapter concentrates on organization and content; Appendix A is devoted to clear writing, applying equally to research proposals and other exposition. Even if no funding or prior approval is required, the rigor demanded by articulating your knowledge of a subject and your research intentions can help you. It can help you to identify inconsistencies in logic and inappropriate fits between the questions you ask, the data you intend to collect, and your methods. Describing what you plan to study, how you will do so, and why it is important is a vital, and in many instances mandatory, precursor to conducting scientific research.

Successful scientific proposals convey good salespersonship: you are selling a future product, your research idea and protocol, to a critical audience that must select from among many such products. To sell your proposal effectively, you must know who your audience is, whether it comprises highly trained specialists in your field, established researchers in related fields, or board members seeking research that is consistent with their foundation's funding directives. Persuasive writing does not compensate for scientific competence, but many good ideas and research designs are undermined by explanations that are either too vague or too specific to the audience evaluating them.

FORMAT

Funding agencies and university graduate programs may differ in their format and content requirements for scientific proposals. A proposal to a conservation or biomedical foundation will require a different emphasis than a proposal to a behavioral society. Before beginning to write, you should obtain all of the available information and forms from all of the funding agencies or programs to which you expect to submit proposals. Books listing funding agencies can be found in the reference sections of most university libraries, and many funding agencies can now be located by searching the Web using general phrases such as "funding animal behavior research." These guides to grants and foundations include the addresses and telephone numbers, or websites and email addresses of funding sources, as well as summaries about the programs. You may be able to download application materials directly from an organization's website, but otherwise you will need to write or call to request application materials, which usually take one to two weeks to arrive.

All federal and many private grants will require approved animal-care protocols. So, simultaneously with soliciting grant information, also begin looking into laws, regulations, and guidelines applicable to use of animals in research. A good place to begin is the unit on your campus that approves animal-care protocols. Furthermore, many professional societies – among them the Society for

Neuroscience, the American Psychological Association, the Animal Behavior Society, and the Ecological Society of America – publish guidelines on animal use and other ethical concerns.

When you have obtained application materials from the granting source, read the instructions carefully, paying particular attention to:

whether proposals can or must be submitted online, or the number of hard copies you will need to submit;
page or word limits;
funding criteria and amounts available;
deadlines;
supporting documents.

Supporting documents are such things as letters of recommendation, academic transcripts, evidence of collaboration, and research permits. It may be useful to create a computer spreadsheet of various deadlines and required supporting documents that you can skim at a glance. There is nothing more frustrating – or impeding to research – than discovering at the last minute, after completing your proposal close to deadline, that you have forgotten something, for example forgetting to obtain a needed document that requires lead time to obtain, or neglecting to allow time to reproduce the required number of copies of a proposal. If any information in the requirements is unclear, be sure to contact the program's funding officer directly. Getting help before submission, either to clarify ambiguous instructions or to guide you as you write, will greatly alleviate unnecessary disappointments.

Here we use the standard Doctoral Dissertation Improvement Grant proposal's format of the NSF as our model for preparing proposals. Not only is the NSF one of the primary sources of funding in the USA for behavioral biology and related disciplines, but also it requires one of the most detailed proposal contents. Learning how to write a competitive NSF proposal will help you in other fundraising efforts because, once completed, the proposal can be adapted to other funding sources.

Unlike scientific journals, which expect to be the sole evaluators of a submitted manuscript, most funding agencies encourage multiple submissions to various sources. They ask you to state whatever sources you have applied to in the event of any overlap in budgeted items. For example, only certain programs at the NSF (e.g. animal behavior, ecology, physical anthropology) have the Doctoral Dissertation Improvement Grants for field studies, but many agencies and even some private foundations have similar programs. For instance, the National Institutes of Health (NIH) has programs for predoctoral fellowships.

Whatever a grant is being applied for, follow precisely all the instructions, including those for specific format. A very important set of instructions is that relating to the length and format of sections, especially the heart of the proposal, where you explain hypotheses to be tested and methods to be used. Federal granting agencies in particular may specify not only a page limit but also the minimum size of margins and the size of fonts for the text. In general, a single-spaced proposal is best formatted in a 12-point font whereas a double-spaced proposal can be put in 10-point. Remember that reviewers are inevitably overworked people who must read a number of proposals in a short period; do everything possible to make your proposal easy to read.

CRITERIA FOR EVALUATION

Each funding agency has its own criteria for evaluating proposals. Scrutinizing these criteria before you write or adapt your proposal will help you to avoid off-target or inappropriate submissions. Some funding agencies provide written evaluations when they announce their results; others will discuss the reasons for a negative decision if they are asked, and still others provide little or no feedback even on request.

The cover page of an NSF proposal asks you to identify the area under which your proposal falls. When it is processed, the proposal will be sent to the program officer in that area, who will ask multiple reviewers to evaluate your proposal anonymously. These reviewers receive a copy of your proposal, which they are expected to treat as a confidential document, and a list of criteria for evaluating your proposal.

The criteria used by reviewers may change from submission to submission, so it is wise to verify what the current criteria will be. Many foundations and agencies make available to investigators printed information concerning how their proposals will be reviewed. The following criteria come from the NSF's Doctoral Dissertation Improvement Grants (DDIG) in the Directorate for Biological Sciences, available at www.nsf.gov/pubs/2002/nsf02173/nsf02173.htm. At the time the first edition of this book went to press, the NSF directed reviewers to consider four items in their evaluation: research performance competence, intrinsic merit of the research, utility or relevance of the research, and effect of research on the infrastructure of science and engineering. Since then, the NSF has revised its criteria down to the following two items, accompanied by the questions that the NSF advises investigators to address and reviewers to evaluate.

What is the intellectual merit of the proposed activity?

How important is the proposed activity to advancing knowledge and understanding within its own field or across different fields? How well qualified is the proposer (individual or team) to conduct the project? (If appropriate, the reviewer will comment on the quality of the prior work.) To what extent does the proposed activity suggest and explore creative and original concepts? How well conceived and organized is the proposed activity? Is there sufficient access to resources?

This criterion evaluates the heart of the proposal: how the research proposed fits into and extends or clarifies existing knowledge in the field. It encompasses the background and justification for the research, the general objectives and specific hypotheses, and the ways in which data will be interpreted.

Is the investigator capable of carrying out the proposed research, and is the approach that will be taken technically sound? Evidence of pilot studies and preliminary results will be especially useful to help beginning researchers establish their credibility and capability, and should be included whenever possible. The technical soundness of an approach will be apparent if the study involves well-established methods of data collection and analyses. Nonetheless, it is critical to select methods appropriate to the kinds of data that will be needed to address the study's goals. A mismatch between questions and methods can undermine a proposal's credibility and will raise questions about the likelihood that the research will ultimately contribute new information to the field. If sophisticated methods will be employed, then references or letters of collaboration from appropriate laboratories or investigators will help to justify your rationale. Application or development of new techniques will require careful and detailed documentation of their suitability and viability. Often the most difficult, and potentially devastating, feature of a proposal is the use of new techniques that reviewers are unfamiliar with or skeptical about. Although as of 2005 the NSF seeks evidence of creative and original approaches, it is imperative to make a solid case for how your proposal will provide new insights to address an important question in your field.

Research in foreign places requires some special demonstrations of feasibility. It is almost imperative that the researcher has had previous experience at the field site and can demonstrate that the proposed methods will work. Local monies and small grants from private foundations should be sought for support of pilot studies abroad. Furthermore, the applicant's plans are more likely to get sympathetic attention if he or she can state fluency in the local language where field work will be done.

Laboratory research also requires evidence that sufficient infrastructure and resources will be available. It is, therefore, important to demonstrate that you will have access to an adequate number of subjects under the appropriate conditions for the study you propose, and that essential equipment and expertise will be accessible. An application that anticipates obvious questions such as these will be more persuasive than one that leaves reviewers doubtful about the proposed study's feasibility.

What are the broader impacts of the proposed activity?

> How well does the activity advance discovery and understanding while promoting teaching, training, and learning? How well does the proposed activity broaden the participation of underrepresented groups (e.g. gender, ethnicity, disability, geographic, etc.)? To what extent will it enhance the infrastructure for research and education, such as facilities, instrumentation, networks, and partnerships? Will the results be disseminated broadly to enhance scientific and technological understanding? What may be the benefits of the proposed activity to society?

Here, the reviewers will be looking for the broader impact of a study beyond its immediate significance in a particular or related set of scientific fields. Broader impacts include contributions that the research will make to society at large, whether through education, increasing diversity of participants, or developing new technologies. Whereas behavioral studies may lead only rarely to new technological developments, they can (and should) nonetheless be relevant to larger questions. Imagine trying to justify the importance of a study on bird territoriality or monkey feeding ecology to an economist or political scientist, or even your grandmother. Think broadly here, perhaps in terms of understanding aggression or seasonal nutritional stress, to identify and "contextualize" your study within a larger more comparative framework. Many discoveries from basic animal behavior research can have great relevance for addressing conservation concerns or issues in biomedical and behavioral sciences. Interdisciplinary studies or integrated field and laboratory approaches can be similarly justified, and any study that will involve the training of assistants at any level should be explicit about the efforts that will be made to extend opportunities to diverse participants.

FEEDBACK

Established scientists with high funding success routinely solicit feedback on drafts of their proposals from colleagues. It is even more unrealistic to expect a beginning scientist to be able to write a competitive proposal without extensive guidance and feedback before submission. Indeed, NSF Doctoral Dissertation Improvement Grants (and many others) require that the student's faculty advisor sign as the principal investigator, indicating his or her endorsement of the proposed research.

Most faculty will not want to approve a study or a proposal unless they are satisfied with it. Therefore, even if they agree with the research topic, they will nonetheless expect to play an active role in shaping the proposal. Regular meetings should be established between you and your advisor, who should read your proposal and provide feedback, and reread revisions until the proposal is ready for submission. Working on a proposal in parts enables you to fine-tune it before investing the time and energy into a full-length draft. It may be difficult for an advisor to provide such specific feedback until he or she sees the entire proposal in writing. It is almost always the case that even then, this draft will undergo multiple revisions. As noted in Appendix A about writing in general, the rule is revise, revise, revise!

Extensive revising is a process that will help you to fine-tune your ideas and allow you to be more explicit and unambiguous in your thinking as well as your writing. Working on your grant proposal with your advisor also immerses you in the discourse of science.

It may be helpful to establish a schedule for completing sections of a proposal. Whereas an advisor should be the primary source of feedback, other committee members or colleagues should be sought for advice and feedback as the draft develops. Such feedback is especially important if letters of recommendation from faculty other than or in addition to the advisor are required.

It is important to remember that funding is highly competitive and many highly qualified, highly rated proposals nonetheless fall below funding limits. The NSF asks reviewers to rate proposals as *excellent* if they are of the highest funding priority; *very good* if they should be supported; *good* if they are worthy of support and sufficient funds are available; *fair* if they are interesting but problematic; and *poor* if there are serious deficiencies. Many very good and even excellent proposals fail to achieve funding status. In these cases, it is worth contacting the program officer to discuss whether a revised submission is merited. Fair and poor proposals usually require extensive revisions to the point that a subsequent submission may bear little if any resemblance to the original.

Rejections or unsuccessful proposals are not uncommon, even among highly respected, productive, and established scientists. It is important to take reviewers' comments and criticisms seriously, but also not to let a negative review or outcome discourage you to the point of despair. Discussing your reviews – and how to improve on or modify the proposal and the research – with your faculty advisor(s) is an important component of learning how to carry out a scientific study and write a successful proposal.

Often, the most rate-limiting steps to proposal writing are the turnaround times between feedback and revisions. A proposal that is the length and depth of an NSF submission may take months of formulating and rewriting; sufficient lead time and coordination between student and advisor should be allowed before submission deadlines.

PROPOSAL CONTENT

This section is subdivided to reflect the different components of an NSF proposal. It may be helpful to read through the entire section before beginning to write, and then return to each topic as you reach that part of your proposal. The generic protocol described here will obviously require some modifications to fit your own research area. There are differences between field- and laboratory-based behavioral research, between observational and experimental research, and between the kinds of questions and methods that can be applied to different species. Be sure to take these differences into account when using these suggestions to guide your own proposal.

Significance of a title

The title of your proposal should state precisely what your study is about. Many first-time proposal writers underestimate the importance of an informative title. If you cannot state concisely what you intend to study, then you may need to clarify your thinking on the subject.

One of the most common problems with titles is that they claim to cover more than the research will actually achieve. For example, a proposal titled "The effects of nutrition on chimpanzee reproduction" implies that nutritional intake and reproductive condition will be measured and correlated. Suppose the study actually involved the use of feeding observations without corresponding nutritional analyses and focused exclusively on female reproductive status. A more appropriate title to the research would then be "The effects of *food (or diet)* on reproduction in *female* chimpanzees," or perhaps more accurate to an observational study, and therefore better, "The dietary *correlates* of reproduction in *female* chimpanzees."

At the same time, however, a title should not be defined so narrowly that it fails to capture the broader scientific context in which the research is situated. Correlations between variables other than diet and reproduction, such as dominance rank or age, might constitute equally important questions in the proposed research. Note the different emphases, and what each implies about the research goals, in the following possible titles:

"The effects of diet on age, rank, and reproduction in female chimpanzees"
"The effects of reproduction on diet, age, and rank in female chimpanzees"
"The relationships among diet, rank, and reproduction in female chimpanzees"

Proposals rarely fail solely due to an inappropriate or misleading title, but like all first impressions, titles set up expectations for what will follow. An accurate and informative title will help ensure that your proposal is sent to appropriate reviewers and that they will not need to readjust their expectations once they begin to read your proposal.

Identifying objectives

What are the general goals of your study? Usually, these objectives are to examine, expand, investigate, explore, develop, or evaluate a set of data relevant to a set of questions that inform the overall research. Objectives should be concise statements that provide enough detail to communicate the scientific focus of the study.

In defining your objectives, it is helpful to think in terms of three or four broad aims. These may be parallel or organized along a gradient from specific to general. A study on "Sex differences in territorial behavior in a polygynous bird (species)" might have three objectives: (1) to evaluate the presence and degree of sex differences in territorial defense; (2) to explore the reproductive correlates of territorial defense for males and females; and (3) to develop a model of the dynamics of polygyny. Remember, however, that you will need to elaborate on each of your objectives in the body of your proposal. Therefore, when formulating your objectives, as in the case of your title, be careful that you do not set yourself an objective that your research cannot address. Objectives, like titles, may require fine-tuning as you develop the body of your proposal.

Integrating your research with existing knowledge

The background to your proposal provides the formal scientific context from which your study is derived and to which your research will ultimately contribute. The background is a section in which you review what is already known and what the outstanding questions in your study area are. This section is, in essence, a formal review that should outline, usually in the third person, what stimulated your interest in your proposed research. The section cannot be written until you have a thorough command of the literature in your field and have identified existing questions or gaps in this literature. How you focus this section – what you choose to include or omit – will depend on what your study proposes to accomplish. The section is often the first part of the body of a proposal, but in relating your stated objectives to existing knowledge you may find it helpful to begin by developing separate background sections for each of your objectives.

If one of your objectives is to evaluate an existing model with data from a new species, then you will need to review the model and provide evidence for why such a test of the model is important. Exceptions that challenge the model or paradoxical features of your study subject are examples of how you may be able to situate your research contribution within a broader context. Similarly, if one objective is to document the existence and degree of sex differences in a particular behavior, then you will need to describe why you might or might not expect sex differences to occur using examples from other studies that demonstrated inconsistent patterns.

It is important to recognize that your preparatory reading is likely to be much more extensive than what you will have space to review in the background section of your proposal. It is also the case that not all of your prior reading will be equally relevant to your proposed research. Choose your examples and citations carefully, being sure to indicate whether your list is inclusive or selective. Profiling an example as being "the sole exception" when there are others that you fail to acknowledge will raise doubts in reviewers' minds about your ability to integrate your proposed work with existing knowledge. The background section should demonstrate that you have a clear idea of what is new about your proposed research. Inappropriate claims about your own originality may offend reviewers who have done similar work or know of other work in the area, undermining the credibility of your proposal and assessments of your ability to interpret your data if you are given the opportunity to collect them.

Hypotheses and predictions

The questions that you propose to address in your research will be clearer if they are framed in terms of specific hypotheses (models) and predictions. Carefully conceived hypotheses demonstrate that you are aware of how your research fits into prior theoretical or empirical work in your area, and carefully deduced predictions indicate whether your reasoning is logically sound. Examples of research hypotheses were given in Chapter 1 and others will be mentioned here. There are usually at least two alternative hypotheses that could be made for any research question raised. The basis for each set of alternatives should be provided and properly referenced. Alternative hypotheses should encompass all possible outcomes of the inquiry. When possible, they should be mutually exclusive, making different predictions. The proposal should make explicit reference to how your data will enable you to distinguish among alternative hypotheses.

You may have strong reasons to believe that your data will support some hypotheses and predictions more than others, but reviewers will be looking for evidence that you have anticipated the possibility of unexpected results and are prepared to deal with them if they occur. It is legitimate to identify which prediction is most likely to be confirmed, but it is often the juxtaposition between outcomes that conform to expectations and possible exceptions that generates the most original and important results. That is, if you are 99% confident of a particular outcome based on what is already known on the topic, then you may lead a reviewer to ask why he or she should endorse a study with such a certain outcome. On the other hand, if you devise an alternative model that makes the same expected prediction but makes other predictions not generated by the traditional model, then this would be a powerful way of setting up your research.

Each of your objectives should have a set of predictions derived from existing knowledge and reviewed in the background section. If your objectives are arranged hierarchically, you may find it organizationally helpful to treat the background and associated hypotheses and predictions for each objective separately. If your research involves a series of controlled experiments, then it may be simpler to write an integrated background section and elaborate on your alternative hypotheses and predictions when you describe your experimental methods. Be aware that your study's results may lead you to rearrange the way in which you present your objectives or experiments in your thesis or publications (see Chapter 3), but until you actually do the research, you must use what is already known in order to organize the logical development of your hypotheses.

Wherever you situate them, a critical step in formulating alternative hypotheses is to articulate the null hypotheses against which any alternative predictions are compared. In statistical terms, a *null hypothesis* (H_0) usually refers to no difference between two sets of measurements or no relationship between two or more variables, when research hypotheses predict such differences or relationships. The operational statement of the investigator's research hypothesis (or the prediction deduced from it) is called the *alternative statistical hypothesis* (H_1, if there is only one).

Rejecting a statistical null hypothesis such as "no difference between two groups" may permit you to conclude that a difference does exist, but it will not tell you the direction of this relationship automatically. Failure to reject a null hypothesis can be highly informative. For example, larger male mammals may need to feed for longer periods than smaller females in order to sustain their body weight, whereas females may need to feed for longer periods due to the energetic demands of gestation and lactation. The null hypothesis in this case, that there are no sex differences in the length of feeding periods, might indicate that female reproductive requirements offset the requirements of male size. Such a conclusion could be drawn, however, only if the body size differences and reproductive states can be measured and are represented in sufficiently large sample sizes to compare (see Methods).

If alternative hypotheses are not mutually exclusive, then it will be difficult to persuade a reviewer that you will be able to resolve the questions you are trying to address. For example, a set of alternative hypotheses stating that dominant females spend a greater proportion of their time feeding than do subordinates (H_1) and that monogamous females spend a greater proportion of their time feeding than do polygynous females (H_2) may be difficult to confirm or refute if dominant females also tend to be monogamous. Similarly, it is critical that your hypotheses fall within the scope of the data you will obtain. In the foregoing example, it may not be possible for you to ascertain the dominance rank of monogamous females because their monogamy precludes the kinds of repetitive social contests that permit such calculations. Or, it may not be possible to evaluate the effects of monogamy and polygyny on female feeding behavior because you will not have a large enough sample size for each possible case. If you cannot discriminate among all of your alternative outcomes, then restating the question and the supporting hypotheses may be imperative. Once again, if reviewers are not convinced that your database will permit you to evaluate your hypotheses, then they will challenge the feasibility of your research despite strong marks for scientific interest.

Precision in your choice of words when you state your hypotheses and predictions, and in your methods of data collection, will help avoid unintentional mismatches. A null hypothesis stating that female reproductive condition has no effect on female diets can be interpreted in different ways. It could imply that you predict no dietary differences among females, or that non-reproductive variables, such as age or rank, are more directly responsible for dietary differences between females. How will female diets be evaluated? Total feeding minutes, proportion of daylight hours spent feeding, feeding rates, or types of food are all valid estimates of diet, and the data that will be used should be specified in order to avoid unnecessary confusion. Of course, it must also be clear that you will be able to assess differences in female reproductive conditions and that you will have large enough sample sizes to compare.

Methods

How you will obtain the data necessary to evaluate your hypotheses is as important to a reviewer as the pertinence of your questions. Using established methods whenever they are appropriate will facilitate comparisons between your study and other related research and will avoid the necessity of detailed explanations. Standardized methods or experimental techniques should be described and referenced fully. Any deviation or innovative methodology will require detailed explanation, justification, and often evidence that you have tested the suitability of the method. If you have presented preliminary results from a pilot study, either in the background section or in a separate section of your proposal, then you will need to explain any deviations from the methods used previously. New or controversial techniques are unlikely to impress a reviewer unless you can demonstrate convincingly that they work.

Included in a proposal's methods section should be information about your study site or laboratory facilities, the duration of the study, the unit of sampling (days, hours, minutes), and the number of study subjects you will be sampling. It is not sufficient to say that you will collect systematic data on all of the adult males and females in your study group, or on all of the groups in your study area. How many males and females or groups will compose your sample set? You may not know the precise answer to this question if you lack prior experience or information about your study population or area. Nonetheless, you can indicate what you will consider to be a minimal sample size, and be prepared to justify why you are confident that your sample size criterion can be met.

You will need to work closely with your advisor and other experienced scientists in order to establish appropriate sample size requirements that will be possible to meet under your particular research conditions and that will be sufficient to address your research questions. One of the most common dilemmas that we have encountered in our own research and advising experience arises from the need to choose between sampling many different individuals versus sampling few individuals, but each more intensively. If too few different individuals are included in a study, it will be difficult to evaluate how much variability exists among them. Conversely, if so many individuals are included that none can be sampled with adequate frequency, it will be difficult to establish individual patterns.

Furthermore, we recommend that you consult statistics books or advisors to confirm the sample sizes that you will need for the particular analyses you intend to use. Experimental design is a distinct sub-discipline within statistics, often with entire university courses being devoted to the subject. Although this topic is actually just the quantitative aspects of experimental design, it is critical to good research planning. Consider a very simple but realistic example. You have reason to test the prediction that a sex ratio in some population of organisms is unbalanced. No matter what the outcome, a sample of five individuals cannot establish two-tailed statistical significance of a difference; indeed, seven individuals of one sex and one of the other is similarly not significant by a binomial test. Considering the natural variation in any biological system, sample sizes needed to evaluate a phenomenon with statistical reliability are often much larger than beginners in science realize. And if the phenomenon under study is complex, requiring multivariate statistical evaluation such as analysis of variance, then the minimally adequate sample size could be enormous.

One effective way to demonstrate your awareness of the importance of obtaining adequate sample sizes is to include a statement about the statistical power that your sample is likely to yield. Consult statistical references or experts to determine the minimum sample sizes you will need for the kinds of analyses you intend to conduct, and then explain in your proposal how you will meet these requirements.

A second and, in some cases, equally important concern is that of pseudo-replication, essentially treating the same data as if they were independent. Problems with pseudo-replication can arise, for example, when the same individuals are sampled under multiple conditions without statistical controls for intra-individual variation.

Again, it is best to anticipate and respond to questions that reviewers may have about your sample sizes and analyses.

Remember that you will need to explain how you will obtain data on each variable you mention in your hypotheses. Predictions from those hypotheses need to be stated explicitly in terms of data to be collected. Definitions of categories of different food types or different types of aggressive behavior, for example, should be stated precisely. A detailed ethogram or coding system can be mentioned in the methods section with reference to a complete listing provided in an appendix. If you also intend to examine seasonal effects on one or more variables, you will need to describe how these effects will be measured and, if appropriate, what type of equipment you will use in order to obtain these measurements.

Consider some further questions. Will you be working alone or with assistants? How will you control for interobserver reliability if more than one person will be collecting data? Will you require preparation time to cut and map a trail system or to habituate your study subjects? How will individuals be marked or identified? Anticipating these logistical considerations in your methods section will indicate that you are fully prepared to conduct your proposed study. Failure to describe these details will raise questions in a reviewer's mind about whether you have a realistic assessment of what will be involved in your study, and it may negatively affect an assessment of your study's feasibility.

The methods section is also the place to describe how your data will be analyzed. Reference to particular software programs and statistical analyses is usually sufficient, but any complicated or unusual analyses should be discussed in greater detail.

Include an agenda, or schedule of research, listing when or in what sequence each phase of your project will occur. A schedule of research will permit reviewers to evaluate whether you have budgeted sufficient time to complete your proposed research. It may be the final subsection of your methods section or a separate section following methods.

Significance

The body of your proposal (background and hypotheses) will have already explained the relevance of your study to your main objectives. The significance section provides an opportunity to flag additional contributions that your study may make. Depending on your research, these might range from the importance of your study to conservation of an endangered species or ecosystem, to advances in technology that your methods will provide, to a multifaceted approach that will help to merge interdisciplinary fields. The significance section is usually no more than two paragraphs, but despite its brevity it demonstrates that you have considered the broader implications of your study to science at large. A good way to think about the significance of your study is to imagine another proposal of comparable merit to your own and explain why your choice of study subjects or location or methods distinguishes yours. The significance section should also include some compelling statements about the broader impacts of your study. Keep in mind here the specific priorities or criteria of the funding organization. The NSF, for example, is explicitly interested in the integration of research and education and in promoting diversity among scientists. Similarly, many conservation-oriented funding agencies will be interested in the impact of your findings on the conservation of your study species or in the region you will work, and in whether you plan to provide training opportunities for local residents or students.

Literature cited

Unlike journals and books, scientific proposal guidelines rarely specify the format for bibliographic material. Following the standardized format described in Chapter 3 will usually be adequate. Literature cited in a scientific proposal is precisely that: an alphabetical listing of the references you cited in your proposal. Careful cross-checking between citations and references is as important in a proposal as it is in a manuscript. (One way of doing this checking by using computers is explained in Chapter 3.) Be sure to take into account any page limits to this section when you are deciding how extensively you will cite existing literature in your proposal.

Summary

The summary of a scientific proposal is the equivalent of the abstract of a research report (Chapter 3), but there is a subtle difference. Proposal summaries become part of public archives and are sometimes scrutinized by legislators or other non-scientists, a point discussed later. Many funding agencies provide a separate page with a defined space for your summary. Other agencies, such as the NSF, simply indicate a word limit. The summary will follow immediately after the cover page of an NSF proposal and, thus, is the first part of your proposal that a reviewer will read.

A proposal summary encapsulates what you intend to accomplish, over what duration, and where. Like the abstract of a manuscript, it should be written only after you have finalized the body of your proposal. Repeating important sentences taken directly from your proposal is a legitimate way to structure your summary, but you will probably need to go back and edit some of the sentences so that they make sense in this context.

Because of potential non-technical readers, proposal summaries need to be prepared with special care. Try to avoid saying things that might seem silly or trivial to a general reader. A US senator once publicly ridiculed a funded proposal in the social sciences because it seemed from the summary to be about the game of tennis. In fact, it was a serious study of frustration and anger as expressed in a defined social context, so it proposed to use observations of tennis players as subjects. Perhaps more careful wording of the summary would have helped prevent the undeserved public ridicule. Studies of animals are especially prone to misunderstanding, as they often seem to the non-scientist as somehow less scientific than studies of brain function, genetics, or ecosystem dynamics.

Budget and budget justification

NSF proposals have a separate budget page included in their application materials. Download and photocopy this page from the application pamphlet to use as a worksheet, and follow carefully the instructions for completing it. The main categories in a budget include salaries for research personnel; research-related travel; permanent equipment; materials and supplies; and indirect costs. You will need to assess carefully the expenses associated with your research and provide an itemized explanation of costs within each category in an accompanying budget justification. The NSF Doctoral Dissertation Improvement Grants have specified caps ($12 000 in 2005) and do not include indirect costs (discussed later). It is important to verify (with the appropriate program director) whether a new limit has been set. Regular research grants do not have formal budget limits, but funding agencies often cut a requested budget, so it is important to think about which items might be expendable if such cuts are recommended.

Salaries must be broken down by the number of paid personnel and their percentage contribution to the research. Graduate students rarely receive salary to conduct their dissertation research under an NSF Doctoral Dissertation Improvement Grant, and faculty advisors never receive salary for their advising role in a student's research. Be sure that you are informed fully about associated costs for any legitimate salaries requested. In some countries, for example, registered employees are legally entitled to vacation wages; most university employees are entitled to fringe benefits, calculated as a percentage of their time contribution to the research and enforced by university research administrations. The lag time between writing your proposal's budget and when you will actually begin your research may mean that you will have to factor in an estimated raise for your personnel for the following year. Many first-time proposal writers are unaware of the various components that determine the real cost of personnel. These costs can add up quickly, approaching the maximum funding obtainable from an agency. It is wise to be conservative and request salaries or partial salaries only when absolutely necessary.

Research-related travel can include air or other modes of travel between your home base and your field site or laboratory facility. It can also include travel costs associated with supply trips. Round-trip Apex airfares can be obtained from any travel agency. Federal funding agencies have strict requirements dictating the use of national air carriers, even if they are more expensive than foreign airlines. Be sure you obtain the appropriate price estimate to include in your budget.

Permanent equipment is defined as any single item costing more than some criterion amount. Equipment might include such essential things as computers for data collection or analyses, or a vehicle. Universities may insist that such equipment be purchased through their own vendors, so university purchasing departments should be consulted for price listings and bidding procedures. Also be aware that the dollar criterion for permanent equipment fixed by a granting agency may differ from the definition of capital equipment used by your institution, and both kinds of criterion values are changed from time to time (mainly to keep pace with inflation). Knowing your institution's criterion for capital equipment may become critical later, as it usually determines which purchases must be put out for competitive bidding to at least three different potential vendors. Even if an exact item is written into your budget and the proposal is both approved by your institution and funded by the agency, you may later have to endure a bidding process that typically requires 90 days.

Materials and supplies include all other expenses associated with your research. Subsistence costs, excess baggage, field or laboratory equipment, and analysis costs are usually included. It is legitimate to include any materials or supplies that you need for your research that you would not otherwise require. You will need to obtain current prices for each supply you intend to itemize (e.g. Nikon 8×40 binoculars, $xxx; 1000 50-ml polypropylene collecting vials, $xxx) and estimate those that you will lump together in your budget justification (miscellaneous supplies, including plastic bags, aluminum nails, paper toweling, etc.).

Indirect costs (commonly referred to as overhead) are fixed by each university to cover the administrative costs of your research if it is funded. Public funding agencies, like the NSF, will send your grant monies to your university, and you will need to submit requests for these funds as your needs arise. Private funding agencies may be exempt from paying indirect costs and may be willing to send you a personal check. Be careful of such income, however, because it may be reported to the Internal Revenue Service (IRS) as taxable income and require itemized deductions. Indirect costs are usually calculated as some percentage of your total budget request and added to the final project costs. You should consult with your university's accounting office or research administration to determine whether you will need to include indirect costs in your budget and, if so, what they are.

To help guide you in your first budget and budget justification attempt, we include an example of a $12 000 NSF Doctoral Dissertation Improvement Grant budget for overseas research. A total of $2290 was requested for travel, $4210 for materials and supplies, and $5500 for other direct costs. The travel justification was as follows: $2000 to pay for part of a round-trip Apex airfare; $190 to pay for two trips (at $50 each for transport and $45 each for lodging) to consult with colleagues; and $100 for five supply and administrative trips between the field site and the nearest town. The budget for materials and supplies included field notebooks, collecting vials, binoculars and photographic equipment, plant press, metric scale, and plastic storage bags. Each item was listed with quantity and price in the budget justification. The other direct costs covered food and lodging, indicated in the justification at the per diem rate.

Other funding

The NSF and most other funding agencies will ask you to provide information about any other sources of funding you have already secured or have requested (or intend to request) from other sources.

Only current, pending, and intended submissions should be listed. You will also need to indicate whether any of these proposals have overlapping budgets, and how you plan to deal with the possibility of duplicate funds. You may have included airfare in your NSF budget and also requested airfare from a departmental travel fund. If both requests are successful, then it may be possible to rebudget one airfare into another category. In the worst case, you lose the second airfare, which presumably you did not need anyway.

Appendixes

Documentation to support your proposal, including copies of research or collecting permits, collaborations, access to laboratory facilities, and samples of your data-coding system or check sheets, can be included in the appendixes. Appendixes do not replace or supplement the text of your proposal, and some funding agencies have strict restrictions on content or prohibit appendixes altogether. Appendixes may be removed before your proposal is sent off for review, so it is important that you include all essential information in the body of your proposal.

Most universities nowadays require approval of any research involving animals and will not forward or administer a proposal or study that has not been approved by an animal care committee. Find out well in advance from your animal care committee the laws and regulations applicable to your proposed study. For example, the US Department of Agriculture sets regulations on care of captive animals in the USA, and many individual states have further laws. The US Fish and Wildlife Service issues collecting permits for wild animals, and the Bird Banding Laboratory in the US Geological Service controls the banding program for birds. Many professional societies issue their own guidelines for use of animals in research (for example, Dawkins and Gosling 1992). When you submit an acceptable experimental protocol, your animal care committee will supply a letter stating that your research conforms to national standards for animal welfare. You will need such proof even if your study involves non-manipulative observation of wild animals. Be sure to allow time for your proposal to be processed.

Research in other countries and at some US facilities such as national parks requires permits. You will also need to verify the permit requirements if you will be capturing, collecting, or transporting any biological material (e.g. plant samples, live or dead animals, blood, urine, or feces). International transport of biological materials may be subject to customs inspections before departing from a host country and when you re-enter the USA. Check with the US Department of Agriculture, the Centers for Disease Control, and all other potentially relevant agencies concerning import restrictions, and with the equivalent government agency in your host country. You may also need separate permits if you will be working with endangered species or transporting any of their products or by-products. Check with the Convention on International Trade in Endangered Species of Wild Fauna and Flora (CITES) to determine how your study subject is classified and whether permits issued from CITES for your research are required. It is also wise to verify with your airline whether transporting preservatives such as liquid nitrogen are permitted, and what kinds of special packaging and documentation may be required at check-in and baggage claim. You may not be able to apply for research permits in other countries until you have proof of your funding sources. A statement to this effect in an appendix will demonstrate that you are thinking and planning ahead.

Table of contents

The NSF requires you to submit a table of contents indicating where each of the major items is located in your proposal. It is usually the third page (p. iii) of your proposal's front material, following the cover page (p. i) and the summary (p. ii). The guidelines specify the sequence and page limits for each component of the proposal. The NSF requests a curriculum vitae (c.v.), limited to no more than two pages, for each individual involved in the project. For Doctoral Dissertation Improvement Grants, this usually means that you and your advisor must prepare condensed versions of your c.v.s (Tips on writing a c.v. are covered in Chapter 5.)

You will not be able to complete your table of contents until you have completed the final version of your proposal and paginated it. Subheadings within the body of your proposal should be listed even if more than one subheading occurs on the same page.

THE SUBMISSION PROCESS

Once you have completed your proposal and your advisor has approved it, you will need to duplicate the required number of copies required by the various university offices that approve your budget and will administer the funds. Be sure to allow sufficient time to obtain these approvals, which can take up to several days. You will also want to keep a hard copy for yourself and provide one for your advisor. NSF proposals are now submitted entirely online, but you may still need to send one hard copy with original signatures on the cover sheet. Other funding agencies may require that you submit multiple copies of your entire proposal for distribution to reviewers or review panels. Express mail services provide greater security that your proposal will reach its destination by the deadline. Be sure to verify the address and phone number of the agency. Furthermore, you must send to a street address, because express carriers cannot deliver to post-office boxes.

Most funding agencies will send an acknowledgment of receipt of your proposal and any other supporting materials, such as letters of recommendation, within a predetermined time period. Online submissions usually generate an initial automatic response and may also be accompanied by a subsequent confirmation once your proposal has been processed. If you have not heard that your submission has arrived, then it may be worthwhile to telephone the funding agency to verify that it is being processed. Never assume that your submission has been received at its intended destination until you have received confirmation. Most funding agencies have inflexible deadlines and will not accept overdue submissions, so it is important to allow time for any mishaps in the submission process. Most online submission sites include information about how to obtain help should you have difficulty during the submission process.

The NSF reviews may take up to six months; other funding agencies may be faster or slower in providing a response. There is little you can do to speed up the review process for your proposal, and repeated inquiries may unnecessarily irritate program officers. You should, however, notify your program officer of any changes in plans or funding status that occur while your proposal is being considered.

THE REVIEW PROCESS

Most funding agencies will send notification that your proposal has been received. Both the NSF and the NIH send email acknowledgments with the program name and application number that your proposal has been assigned. The program officer responsible for shepherding your proposal through the review process will send your proposal to appropriate reviewers for anonymous evaluations. Reviewers are instructed to treat proposals as confidential documents not for distribution. You may suggest the names of possible reviewers or individuals who may have conflicts of interest regarding your work when you submit your proposal. The program officer, however, will make the ultimate decision about who will review your proposal.

In addition to these ad hoc reviewers, NSF and NIH proposals are also evaluated by two or three members of a panel that meets to consider all requests for funding during the same submission cycle. At these meetings, the *ad hoc* reviews and the reviews by panel members are discussed, and a funding priority decision is reached. Within a few weeks of the panel meeting, the funding decision and anonymous reviews of your proposal will be mailed to you.

Often, if a funding decision is positive, the program officer will contact you to discuss possible or recommended revisions to the budget. You may be asked to submit a revised budget page before your grant can be approved officially. Program officers rarely call applicants with news of a negative funding decision. However, most program officers are willing to discuss your reviews and provide advice about whether a revised proposal should be resubmitted to their panel. The program officer may reiterate concerns raised in the reviews and provide more details on the panel discussion of your proposal that will help you to strengthen a resubmission. Depending on your schedule, the status of other grant applications you may have submitted elsewhere for the same study, and the feedback you receive, you may decide to revise and resubmit your proposal for the next funding cycle or wait to develop a new or substantially different study for possible funding. Even if your proposal is funded, we encourage you to consider carefully the reviewers' comments and incorporate any recommendations that make sense into your actual study.

WHAT TO DO WHILE WAITING FOR A DECISION

Many researchers consider the interim between submission of a proposal and notification of its funding status to be wasted "downtime," but we think that this is an erroneous viewpoint. There are many time-consuming preparations that can be initiated during the waiting period, such as arranging purchase of supplies that require funds once you have obtained them.

If you will be working in another country, use the waiting time for preparations. Make sure that your passport will be valid for the duration of your intended study. You may need to wait until you have secured funding before applying for a research visa, but you can obtain all necessary forms and complete most of them in advance. Some countries require proof that you have been vaccinated against particular diseases; if you will be working with wild animals, you might consider having rabies and (if you have not had a booster in the past 10 years) tetanus vaccinations. Consult your doctor about foreign travel, and make an appointment for a complete medical examination before beginning your research. If you do not already have health insurance, you should obtain it before traveling overseas. If you do have insurance, it is important to verify that the policy will cover you during the period and under the conditions of your research.

Familiarize yourself with the policy's limitations and liabilities so that you are prepared in the event of an emergency. If you will be working in an isolated area, you should also begin to assemble a medical kit that includes a guide to self-administered first aid. The contents of the kit should match the risks of the environment and might include items such as insect repellent, a snakebite kit, antibiotics, and any prescription medication required for allergies.

You may be required to notify your university if you will be off-campus for any length of time. Often this will change your registration status and lower any fees you are required to pay. Obtain the necessary forms in advance so that you can process them as soon as you know your precise departure date.

When you developed your budget, you will have discovered where to purchase the supplies you will need. Assemble order forms for each source, but be sure to verify the prices, including carrier fees, at the time you are ready to make the purchase.

You should also already have a good idea of what methods you will be using in your study, but the waiting period is a good time to practice these methods and fine-tune them as much as possible. Do you know how to use a compass, binoculars, or sound-recording equipment? Borrowing the equipment and testing your use of it on local animals will save you valuable time later on. Do you have adequate experience of mapping or quantifying vegetation? If not, obtain some training or practice now. Similarly, if you have designed a check sheet or coding system for recording behavioral observations, now is the time to familiarize yourself with it. If your study involves observations of wild primates, a combination of practice sampling sessions on the same or a related species in a local zoo or on a locally occurring species, such as squirrels or birds, will help you train for your planned field study. Such training will give you greater confidence when you begin your actual research.

3 How to write a research report

Front matter
    Title page
    Abstract

Body of the report
    Section heads and subdivision
    Introduction
    Methods
    Results
    Discussion
    Series of experiments

End matter
    Acknowledgments
    Appendixes

References
    Literature cited section
    How to cite references

Tables and figures
    Tables
    Figure legends
    Figures
    Types of figures
    How to cite tables and figures

Submission and review
    Friendly pre-review
    Where to submit
    The covering letter
    The review process
    Upon receiving the editorial decision
    How to review a manuscript

There can be no great smoke arise, but there must be some fire, no great report without great suspicion.

John Lyly (c. 1554–1606)

Research results are of little use unless they are reported so that others can access them. We have turned "report" of the opening quote into a double entendre in order to make the point that a scientific report should be written clearly so as to avoid suspicion. This chapter deals with writing a research report such as a graduate thesis or a manuscript for publication. The next chapter considers presenting research orally or by a poster paper at a scientific meeting, and Appendix A discusses clarity of writing in general.

Scientific reports have a standard format that should always be followed in the absence of special instructions to the contrary. Here is the usual sequence of parts of a scientific report:

Title page
Abstract
Introduction
Methods
Results
Discussion
(Conclusions)
Acknowledgments
Literature cited
Appendixes
Tables
Figure legends
Figures

Exceptions to this list are uncommon. For example, certain journals may have unusual requirements of style or form that need to be followed in manuscripts. Our advice to always follow exactly the instructions to authors when preparing a report for publication cannot be overstressed. In certain student reports and theses, placing figure legends on the figures themselves and inserting figures and tables as separate pages after their first mention in the text may be permitted. Conclusions may be a separate section (older pattern) or incorporated at the end of the discussion (now more common). Acknowledgments in some journals are placed in a footnote on the first page. A few journals follow the older format of replacing the abstract with a terminal summary that is inserted just before the literature cited. These are the only common exceptions to the sequence of parts listed here, although see *Series of experiments* later in this chapter. Even in brief articles and short communications that are not divided by subheads, the sequence of material follows the list given. In all cases, however, we recommend that you consult the journal's instructions to authors and consider our comments on each section of a written report in the context of more specific instructions. Each of these parts is now discussed in detail; the chapter concludes with more general notes.

FRONT MATTER

Most journal manuscripts begin with a title page followed by an abstract on a separate page. The body of the paper therefore begins on the third page. The title and abstract need to be crafted carefully, for each needs to be concise and informative to induce the reader to read further.

Title page

Many journals require a manuscript to have a separate title page, although others allow the abstract to appear on the same page. For manuscripts of monographic length, a table of contents may be required on or after the title page. In general, though, a title page consists of the work's title, the name(s) of the author(s), and other information appropriate to the type of report.

The other information on the title page varies with the situation. Student papers often need to bear the course name and number, the name of the faculty supervisor, and the date. Journal manuscripts generally require the institutional affiliation(s) of the author(s) and the mailing address of the corresponding author, including an email address and fax number. Some require a suggested short title to be used as a running head in the journal.

The title itself

As in titles of research proposals (Chapter 2), the title of a report is a tradeoff between conciseness and description of content. The considerations discussed in Chapter 2 apply here, but there are also some minor differences between titles of proposals and reports. It is more frequently the case that one changes the title after drafting the report and may change it again after revision. Journals, like funding agencies, may have specific requirements of titles, such as restriction of their length. At least two types of title found in journal articles might be inappropriate for proposals: full sentences and questions. Here are examples of actual titles from the literature:

"Juvenile experience influences timing of adult river ascent in Atlantic salmon"

"Why is the male wood duck strikingly colorful?"

Some journals may encourage these genres of titles, while others may discourage or even prohibit them.

Journals may require taxonomic identification of the species studied to be included in the title. Often this requirement is merely the Latin binomial of the species (see Appendix A on abbreviations). However, a few journals want more details, such as the family, order, or class of the organism. We believe that common sense should be used in such specifications. For example, birds are very well known, so it is usually unnecessary to include a lot of taxonomic identification.

The author's name

Yes, of course you know your name, so what is there to discuss? Well, it is the bibliographer's nightmare if the same person publishes under different names. So, in your very first manuscript for publication decide upon the name you will use and stick to it forever.

The problem of multiple names of the same person is more common than you might think. Our late colleague John T. Emlen, Jr, published under that name and under J. T. Emlen, Jr, while his father was alive. Then he dropped the junior designation, thus creating two other publishing names. His son John Merritt Emlen published under that name and under John M. Emlen, and then switched to J. Merritt Emlen. So there seemed to be a lot more John Emlens out there than was actually the case. Jane Goodall published first under her maiden name, then under Van Lawick-Goodall after she was married, and then went back to Goodall. Thus, even in the same bibliography, her papers are not listed together, some being in the Gs and others in the Vs. Do not contribute to the confusion: pick a publishing name and stay with it.

Issues of coauthorship

The names that should be included in the authorship of a manuscript are not always obvious. (The same considerations apply to abstracts for scientific meetings, discussed in Chapter 4.) Misunderstandings about coauthorships can cause bad feelings and antagonistic relationships and may even lead to accusations of unethical scientific behavior. Unfortunately, no uniform guidelines exist concerning the specifics of coauthorship that apply across all areas of the biological sciences. One exception, perhaps, is that all authors listed on a manuscript must have read and approved the contents of the submitted document and any subsequent revisions before its publication. Some journals require a declaration to this effect in the covering letter. Apart from this universal rule, it is essential to be sure that you understand the culture of authorship specific to your own advisors and collaborators. Confirm your assumptions of others' expectations about receiving credit as early in the research process as possible. If you have any doubts, ask. Some graduate advisors legitimately expect to be included as the last author of any published paper or meeting abstract that results from the research, particularly if any part of the student's education or research was supported by funds from the advisors' own grants or conducted at the advisors' field site or in his or her laboratory. Others expect an acknowledgment for their help but otherwise encourage their students to publish and present papers independently, even though they were required to be identified as the principal investigator on the student's grant proposal (see Chapter 2).

Collaborative research is an even trickier business to sort out, and it is best to do so well in advance of preparing a publication. In some fields, simply providing a particular reagent or lending a piece of equipment is accompanied by an implicit assumption of coauthorship on any publications resulting from the work for which the reagent or equipment was used. In other cases, an acknowledgment thanking that individual for providing a necessary item is considered sufficient. Other ambiguities arise whenever more than one individual is working on a related project or when individual studies rely on unpublished data collected by others. Often, the individual who writes most or all of the first draft of a manuscript is the first author, others who participated or contributed to the data collection and conceptualization of the study in significant ways are second (and third, etc.) authors, and the advisor, laboratory head, or field-site director is the last author. Nevertheless, different individuals may define "significant contributions" in different ways. Furthermore, authorship sequences that may have been agreed upon at the outset of a study may change if the roles of each individual change during the course of the research. We elaborate on some of these points in Appendix B.

We two authors explicitly discuss coauthorship expectations before embarking on any collaborative research effort, distinguishing between contributions to conceptualizing the study, collecting the data, analyzing the data, and responsibility for financing or otherwise enabling the work. In this rather rigorous scheme, any participant is usually expected to have contributed to at least two of the four

Section heads and subdivision

Journals and books vary widely with respect to the hierarchy of heads for sections and subsections. A commonly used hierarchy is primary heads centered above the text, secondary heads left aligned above the text, and tertiary heads (if used) run in with the text (either indented or left aligned). Heading formats also vary with regard to whether they are in full caps, initial caps of all major words, or initial cap of the first word (and proper nouns) only. A few journals number heads hierarchically, either in place of format distinctions or in addition to them. In the absence of specific instructions, use your own system of headings – but be consistent.

Introduction

The introduction is the first major section of a scientific paper: the guide to what you did and why you did it. Some journals use the head "Introduction," but many leave the initial section un-titled. The first paragraph of the introduction is the most important one because it either succeeds or fails to entice the reader to look at the entire paper. A common error is attempting to say too much in the opening paragraph. Keep it short and interesting. Here is a rule of thumb:

┌─ RULE ──

Always state no later than the last sentence of the opening paragraph what the study is about.

The introduction usually contains the general statement of the problem, the name of the species studied (along with the scientific name, even if already given in the title or abstract), and some indication of why the problem is interesting or important. This justification often requires citation of what is already known about the problem, establishing a niche for the work that will be reported. When appropriate, the introduction should conclude with specific predictions that were tested by the study. These predictions may have to be developed by verbal or mathematical logic from a stated model, especially if the deductive reasoning is complex or long.

The importance of stating why your study was done cannot be overemphasized. Here is a quote from our colleague Charles T. Snowdon, who has been editor of both *Animal Behaviour* and *Journal of Comparative Psychology*: "The two major failings of the articles that I reject are not providing a compelling rationale in the introduction and not providing a sufficiently complete methods section to understand how the study was done."

Methods

The term "methods" is generally taken to include both procedures followed and physical equipment used in the study. If "methods" is equated with procedures *per se,* then the section is sometimes entitled "Methods and materials."

Methods sections are often subdivided, but the subdivisions are not standard because they must be appropriate to each individual report. Here are some of the things that could make up the methods section when applicable to a study:

- Field site or study area, with ecological description if pertinent
- Housing and husbandry of captive animals
- Study population, with marking schemes and so on
- Division of subjects into experimental and control groups
- Field times, including dates and number of hours spent in observational studies
- Experimental design and protocol
- Equipment used to take and analyze data
- Criteria used in identifying and separating recognized entities
- Procedures and protocols followed in taking data
- Statistical methods used in analyzing data

It used to be said that a methods section needs to be so complete and clear that another investigator could replicate your study from the information provided. With limitations of space in modern journals, it seems doubtful that a methods section can be so complete. Nevertheless, set as your goal the writing of methods in such a way that your study could be replicated from them. A laudatory trend among some journals is to publish further details about methods on the journal's website.

The methods section is the Achilles' heel of a manuscript. In a proposal (Chapter 2), it may be obvious that referees will focus on the methods, but the same is true for a manuscript under review by a journal. Inadequately described methods may cast doubt on the quality and validity of your results. If you followed a standard, well-established protocol, or your study area has been described in detail elsewhere, you may find that you can summarize the most salient features or cite the more elaborate reference, provided it is published in a readily accessible source. Citations to unpublished sources, such as Ph. D. dissertations, are not especially helpful to a reviewer or reader who would like to follow up on or confirm a procedure. If you modify established methodology or develop wholly new approaches, however, be certain to explain them in adequate detail.

Results

The findings are the heart of your report, so they should be presented clearly and concisely. Where appropriate, provide the results in figures and tables, which are easier to absorb at a glance than are long explanations. The accompanying text needs to point out what can be seen in the figures and tables but should not repeat the information there. Journals are always strapped for space and cannot allow redundancy among text, figures, and tables. Some journals may also limit the number of tables and figures because they are expensive to prepare for printing.

In general, the results section should contain no discussion or interpretation of the findings: "Just give me facts, ma'am" (as Joe Friday was fond of saying on *Dragnet*). If the introduction to the report is laid out well, the reader will see immediately the import of the data. If predictions were made in the introduction, repeat them as a way of introducing each facet of the results.

Results sections, like methods, may be subdivided appropriately for the study. Often, a study includes two or more experiments, so each of these can be presented in its own subsection. Or, there may be wholly different analyses performed on the same set of data, and these analyses can be presented in separate subsections. Subdividing your results will also help the reader to follow the logic of your analyses, as in moving from general results (e.g. overall diet) to specific topics (e.g. seasonal variation or sex differences). You may find that you do not need to present all of the analyses you have conducted while exploring the nuances of your data. Following the sequence of major questions raised earlier will provide a road map for deciding which results are appropriate to include.

Discussion

Discussion sections as drafted, even by experienced authors, are nearly always too long. A common mistake of inexperienced authors is to attempt to wring from their data by discussion more than the data really can show. In a good study, the data speak for themselves if the introduction and results sections have been crafted well.

Discussion sections generally deal with three matters: (1) an analysis of sources of error in the data, (2) integration with what was previously known, and (3) implications for future study. The first matter is too often overlooked, and yet it becomes a valuable adjunct to the methods section for other investigators who will conduct related studies. In fact, analysis of error sources was classically the complete discussion section of a report. The second matter is one that often gets out of hand. Do not attempt a mini-review of all related literature, but do be specific about how your results extend and modify what was already known. Keep the third matter short; reasonable statements about what might be done in future studies are appreciated, but rambling speculation is not.

Discussion sections, unlike other parts of a written research report, often include more tentative language. For example, definitive statements about what was known without the present study (and already published) on a topic can be made in the introduction, just as statements about what was done can be stated without qualification in the methods section. Interpretations of the results and how they fit into or alter what was previously known may or may not prove correct, however. Be careful to distinguish your own interpretations from those of others and from indisputable facts.

Remember, a short, concise discussion section is the hallmark of a good study. Here is another (admittedly controversial) rule of thumb:

RULE

The discussion section should rarely be the longest section of a report.

If the discussion gets out of hand, revise and condense. Often, much of the survey of what was already known belongs in the introduction. Refer back to the introduction instead of repeating the material.

Series of experiments

There is a special exception to the division of the body of a paper into one section each for introduction, methods, results, and discussion – namely, when a manuscript reports a succession of separate experiments, each following from the results of the previous one. If each experiment in the series uses basically the same methodology, then the standard format may be appropriate, with subdivisions of the results section for the successive studies. When the methods for the several experiments differ, however, many journals allow (and some outright encourage) a division of the manuscript by experiment, each stating its own methods and results as separate subheads. In such cases, a single introduction to and a single discussion of the related experiments are ordinarily preserved as in standard format.

END MATTER

Three elements of textual material generally follow the body of the report: acknowledgments, a list of references cited, and any appendixes. A few journals place acknowledgments within the methods section of the paper. Most reports will not have an appendix, although in certain areas, such as mathematically oriented papers, an appendix (e.g. of derivations of equations) is commonplace. Here we discuss only acknowledgments and appendixes, deferring the important topic of references to its own major section.

Acknowledgments

The acknowledgments section is more important than you might think. Many readers will skip over it, but those who helped with the study surely will read it and others will skim through it to get a feel for the intellectual influences on your study. The titular word can also be spelled with an "e" after the "g" – just go with the journal's preferred form.

Rarely is a scientific endeavor truly the sole product of the author; think through the entire study and remember to thank the people who helped you. Failure to acknowledge those who helped you is almost certain to discourage their help in the future. You should also acknowledge research permissions (say, to work on park or private lands) and sources of financial support. Most granting agencies require such acknowledgment and specify in their award letters how their support should be credited. If you work at a field station or research center, you may be required to cite a contribution number in the acknowledgments; obtain that number from the director. Some field stations and research centers provide the number when your manuscript is complete; others provide it only after a manuscript has been accepted for publication, in which case you should have the requisite sentence written into the manuscript with only the number itself blank.

Appendixes

Most scientific reports will not have an appendix. (By the way, "appendices" is an alternative plural.) An appendix is often used for long mathematical development that is necessary to the paper but would be a tedious distraction for the reader if contained in the main text. Sometimes in ornithological works citing many species, scientific binomials are omitted in the text and collected together in an appendix. Student papers and theses often have appendixes to provide more basic data than would be justified in a journal article but are desirable for evaluation by an advisor or examining committee.

There is another, distinctly different potential use for an appendix, or alternatively a similar short section entitled "Added in press." Sometimes the lag between acceptance and publication is so great that an important variable or conclusion described in the manuscript has changed or a critical paper that must be cited has just appeared in the literature. In such cases, write a concise update citing the section of your manuscript affected, and obtain the editor's permission to insert the new appendix or section.

REFERENCES

References to the literature form the core of scholarship through the augmentation of existing knowledge. It is therefore critically important to provide proper bibliographic details in a consistent, organized manner and to cite literature unambiguously. Here we deal first with how to prepare a terminal bibliography, and then with how to reference its entries from the text of your report.

Literature cited section

In a scientific report, every work referred to in the text should appear in this section, and every reference in this section should be cited somewhere in the text. To emphasize this point, the section is called

"Literature cited" in some journals, instead of "Bibliography" or "References."

Reference form

Journals vary widely in the format of references for the literature cited section. A manuscript prepared for publication should slavishly follow the style conventions of the journal for which the manuscript is intended. There are various standards available (see the works listed in the further reading section at the end of this book), but none enjoys widespread, much less universal, adherence. In the absence of specific instructions, we recommend this simple generic format for journal references:

> Surname, initials, initials surname, and initials surname. Year. Title. Journal, volume: inclusive pages.

For example:

> Wilson, E. O., F. M. Carpenter, and W. L. Brown. 1967. The first Mesozoic ants. Science, 157: 1038–1040.

This format is a sort of compromise among many that are commonly used in journals of ethology and ornithology. Therefore, when later adapting the literature cited to a specific journal format, this form will generally require the fewest changes. Some common variants are: (1) surnames in small caps, (2) surnames of second and subsequent authors preceding their initials, (3) initial caps for species' common names in title (mainly ornithological journals), (4) journal name in *italics* or **boldface**, (5) omission of the comma after the journal name, (6) volume number in **boldface**, (7) using a comma instead of a colon after the volume number, and (8) omitting the space after the colon.

Note two things: First, journals of science almost never use authors' given names (merely initials), although other types of scholarly journals commonly use first names. Second, issue numbers are generally not used because they are unnecessary in journals that are paginated continuously within a volume. If each issue (number) is paginated separately, place the issue number in parentheses after the volume number (before the colon). Again, check the requirements of the journal to which the manuscript will be submitted. At least one journal we know of (in herpetology) requires the inclusion of issue numbers for all references.

Listing of books, chapters in books, newspaper articles, unpublished theses, and various other miscellaneous references varies enormously. In the absence of specific instructions, use a generic format similar to the one given earlier, adapting it for the circumstances. For example:

> Woolfenden, G. E. and J. W. Fitzpatrick. 1991. Florida scrub jay ecology and conservation. In: P. B. D. Lebreton and G. J. M. Hirons (eds.), Bird Population Studies: Relevance to Conservation and Management, pp. 542–565, Oxford University Press, Oxford.

In this example, the article/chapter title uses an initial capital for only the first word, whereas the title of the book in which it appeared uses initial caps throughout. A few journals do not use initial caps through book titles, but rather treat them with the same conventions as titles of articles and chapters in using initial caps only for the first word and any proper nouns. Once again, follow slavishly instructions to authors and check recent issues of the journal to which you will submit for examples of format.

The city of publication of a book has become an especially knotty problem as companies increasingly publish books simultaneously in multiple cities. In the foregoing example, the book itself lists on the title page St Louis, Baltimore, Boston, Chicago, London, Philadelphia, Sydney, and Toronto. A few journals want a citation to include all such cities. More commonly, the first city listed with ''and elsewhere'' suffices. An increasingly common pattern is simply to cite the first city listed and let it go at that. Another pattern is to list the city of the home office of the publisher, which may or may not be the first city listed on the title page. Determine this city by looking at the copyright page (usually the back of the title page); US publishers commonly provide there the home-office address, although many others do not.

If you must cite unpublished theses and dissertations, treat them basically like published references. We think it is a good idea to avoid citing unpublished works unless necessary. These are generally unavailable to your readers and have not had the benefit of the same kind of critical peer review as journal articles or books. To cite an unpublished thesis, use this form in the absence of specific instructions:

> McGowan, K. G. 1987. Social development in young Florida scrub jays (*Aphelocoma c. coerulescens*). Ph. D. dissertation, University of South Florida, Tampa.

Other kinds of unpublished material and information – unpublished data, personal communications, manuscripts, works under review, and so on – are generally not accorded a listing in the reference section. These may be referred to in the text in order to indicate the status of a study or manuscript; for example, Hailman (unpublished data), Snowdon (personal communication), Strier (MS), Darwin (in review). Never cite a manuscript as "in press" unless it has been formally accepted by the editor, in which case it should be listed in the reference section with the journal name, even if the year, volume number, and pages cannot yet be provided.

Computer software now exists (e.g. Endnote™) that formats your literature cited section according to a selected journal style. (If a journal is not included in the huge list of choices, you can enter its format so it will be there for the next manuscript too.) The software can work only if you create a database of references within the software. The software then selects from your database those references cited in your manuscript. Therefore, you may create one database for all of your references, regardless of topic.

We close this section with a reminder to check carefully the instructions to authors for the literature cited format of the journal to which the manuscript will be submitted. Journal formats vary more in the form of bibliographic citation than in any other aspect.

Sequence of references

References are usually listed alphabetically by surname of the senior (first) or sole author, but various complexities arise. Again, journals vary as to the rules they follow, but in the absence of specific instructions follow these guidelines:

- European surnames beginning with the separate word "von" (mainly German and Swedish) or "van" (mainly Dutch) are listed under V if that letter is capitalized (e.g. Van Lawick) but by the name that follows if v is in lower case (e.g. von Frisch is listed under F). The same rule applies for "de" (e.g. de Waal is listed under W). If journals use block capitals for all authors' names, however, we advise you to consult published articles in the same journal to see how compound European surnames are treated.
- Within identical authorship, references are listed chronologically. If two or more references bear the same date, they are designated by lower-case letters following the year. Example:

    Brown, J. L. 1964a. The evolution of diversity in avian territorial systems. Wilson Bull., 6: 160–169.
    Brown, J. L. 1964b. The integration of agonistic behavior in the Steller's jay *Cyanocitta stelleri* (Gmelin). Univ. Calif. Publ. Zool., 60: 223–328.

- If two authors have the same surname, they are listed alphabetically within surname by first initial. If first initials are also the same, proceed to the next initial for alphabetizing. Example:

    Smith, S. M. 1991. The Black-capped Chickadee: Behavioral Ecology and Natural History. Cornell University Press, Ithaca, N. Y. 362 pp.
    Smith, S. T. 1972. Communication and other social behavior in *Parus carolinensis*. Publ. Nuttall Ornithol. Club 11:1–125.

Smith, W. J. 1977. The Behavior of Communicating: An
Ethological Approach. Harvard University Press, Cambridge,
Mass.

- When there are multiple-author references with the same first
  author, references by that author alone come first, followed by
  those with other authors alphabetized by surname of the second
  author, unless different instructions are provided in the journal's
  guide to authors. (When there are references with first and
  second authors the same, but different third authors, then those
  with no third author come first and others are alphabetized by
  third author.) Example:

Marler, P. 1969. Vocalizations of wild chimpanzees. Rec. Adv.
Primatol., 1: 94–100.

Marler, P. and D. Isaac. 1960. Physical analysis of a simple
birdsong as exemplified by the chipping sparrow. Condor, 62:
124–135.

Marler, P. and M. Tamura. 1964. Culturally transmitted
patterns of vocal behavior in sparrows. Science, 146: 1483–1486.

How to cite references

References must be cited in the text by an unambiguous method that
specifies an exact entry in the literature cited section. It is never
permissible in scientific publication to attribute a result or viewpoint
to an authority without a specific reference. This section discusses
issues of citing literature, including multiple-author references, page
specifications, Latin abbreviations sometimes used, and how to
check for a match between text citations and the entries in the
terminal bibliography.

Text citation of references

Journals differ a little as to form of citation for references in the text.
For single-author references, citations generally take one of these
forms, depending upon context:

Wilson (1974a,b) showed that . . .

. . . may now be considered conclusively demonstrated (Wilson 1974a, b).

A few journals repeat the year (e.g. 1974a, 1974b), and some insert a comma between author and year for citations wholly within parentheses.

Repeated successive citations of the same publication can be shortcut by use of Latin abbreviations (see later subsection). Not all journals allow these, however.

When two authors cited in the same report have the same surname, a good practice followed by many journals is to include the initials: S. M. Smith (1991), S. T. Smith (1972). Other journals include initials only if the year is also the same, such that without the initials the reference would be ambiguous. Including initials in all cases, however, helps the reader to locate the reference in the literature cited section, and we recommend the practice in the absence of instructions to the contrary.

It is obvious that historical references (author no longer living) should always be cited in the past tense, but for more recent references one might use either the present or past. The present tense – for example, "Snowdon (1990) shows that . . ." – is logical because the reference exists in the present. The past tense – for example, "Snowdon (1990) showed that . . ." – is equally justifiable because all references cited were actually written in the past. We authors think it is good practice to cite references consistently in the past tense in order to avoid the appearance of inadvertently attributing to an author a present view on the basis of his or her past publications. For example, "Marler (1955) believes that . . ." seems to report Marler's current views, whereas those views may be wholly unknown to the writer. Nevertheless, if a reference is very recent and clearly the author's latest contribution to the topic, then it may be more natural to write of it in the present tense.

Multiple authors

Most journals cite both of two authors (e.g. "Washburn and DeVore 1961") but cite three or more authors as "Tinbergen *et al.* (1967)." Note that "*et al.*" is an abbreviation for the Latin *et alii* ("and others"), and so the "*al.*" part demands a period. A few journals include all the authors when there are three, reserving the "*et al.*" designation for four or more authors. At least one journal we know of includes all three authors in the first citation of that reference, and then uses "*et al.*" in every subsequent citation.

A special problem arises when the same first author published two (or more) papers with at least two other (but different) authors in the same year. Suppose these two papers appeared in the literature cited section (as has actually happened):

> Ficken, M. S., R. W. Ficken, and S. R. Witkin. 1978. Vocal repertoire of the black-capped chickadee. Auk, 95: 34–48.
> Ficken, M. S., J. P. Hailman, and R. W. Ficken. 1978. A model of repetitive behaviour illustrated by chickadee calling. Anim. Behav., 26: 630–631.

Some journals handle this situation awkwardly by designating the two as Ficken *et al.* 1978a and 1978b. The problem is that the references may not be adjacent to one another in the literature cited section. We think it better, in the absence of instructions to the contrary, to cite these by full authorship: "Ficken, Ficken, and Witkin (1978)" and "Ficken, Hailman, and Ficken (1978)," but when doing so it would be wise to explain this unusual situation to the editor in your covering letter.

Another tricky problem returns to the issue of different authors with the same surname. In Susan M. Smith's book *The Black-capped Chickadee*, there are nine other Smiths apart from the author who are cited as sole or first authors of references (and likely a lot of other Smiths who are second or subsequent authors). In all, there are 32 references listed that begin with Smith. Three of these would ordinarily be cited "Smith *et al.*" because they have three authors. A journal that allows citation of initials only if dates are identical does the reader a great disservice. In this example, "Smith *et al.* (1986)" is technically unambiguous because the other "Smith *et al.*" references do not bear this date. Nevertheless, the reader does not know where to find this particular "Smith *et al.* (1986)" reference in the 32 papers and books written by ten different Smiths. If the reference is cited as "J. N. M. Smith *et al.* (1986)," it can be located in the bibliography immediately.

The author may want to alert the journal editor to possible confusions that could arise concerning citation of multiply authored papers in the manuscript. Some well-heeled journals, especially in biomedical fields, have paid editors and professional copy-editors who will be aware of most citation problems. By contrast, the vast majority of journals in evolutionary, ecological, and behavioral biology have unpaid editors who are academics from the discipline and who are donating their services. These editors may have limited support, such as a part-time secretary or student hourly help. It is not out of line for an author to alert such editors to citation problems so that the editor can inform the assistants.

If you use reference-formatting software (such as Endnote), most types also format the citations in the text of the manuscript. Here is how Endnote works, as an example; other software may work slightly differently. You compose your text on the computer with your software's database open (next to the word-processor display is most convenient). When you want to cite a reference, find it in the database and drag it to your text in the word processor. Endnote has a special provision for use with Microsoft's Word ™ that makes insertion of a reference even easier.

The software inserts a placeholder for that specific reference. The placeholder in Endnote is enclosed in curly brackets and contains a number, the last name of the senior author, and the year of publication, like this: {427 Strier 2004}. The number was assigned by the software when you entered the reference into the database, and each reference has a different permanent accession number. When you finalize your manuscript and tell the reference software what journal format to use, it not only compiles your literature cited section based on text citations and puts the references in the correct format for the journal but also goes through your manuscript and replaces the placeholders with the citation form specified by the journal.

Page citations

A laudatory practice not allowed in some journals (for imagined want of space) is a page citation with the reference. Page citations are especially useful to locate specific points and direct quotes. The citations usually take one of two forms: either (Hailman 1990: 23) or Hailman 1990, p. 23). Always provide page citations when material is quoted directly.

Latin phrases

Reference citation commonly employs three Latin abbreviations (*ibid.*, *op. cit.*, and *loc. cit.*) in addition to *et al.* (discussed earlier); Latin used in general writing (e.g., i.e., cf., sic, viz.) is discussed in Appendix A. *Ibid.*, *op. cit.*, and *loc. cit.* are used more commonly in the humanities than in the sciences. Even if you do not use the Latin yourself, however, every author should know what these abbreviations mean so as to interpret them correctly when encountering them during reading.

Ibid. is an abbreviation for the Latin adjective *ibidem* (in the same place). This designation is most commonly used in footnotes and bibliographies that are sequential by every citation. That is, if footnote X gives the reference and footnote X+1 refers to the same reference, then *ibid.* is the sole entry of that latter footnote. *Ibid.* is rare in scientific reports because literature cited sections are not usually of the appropriate format, and even when they are a different citation method is used. For example, in the journal *Science*, the references are listed sequentially by first citation, but they are numbered in the literature cited section. References are cited not by footnote numbers but rather by bibliographic numbers. Therefore, if you cite reference X and then cite it again, even if it is the very next reference cited, you refer to it by the same number X.

Op. cit. and *loc. cit.* are similar to *ibid. Op. cit.* is an abbreviation for *opere citato* (in the work cited) and is often used instead of repeating the year in order to help the reader know that the same reference is being made (as opposed to another paper by the same author, for example). *Loc. cit.* is more specific: *loco citato* (in the place cited) refers to the very same passage.

Checking citations

Every reference cited must of course appear in the literature cited section, and vice versa. Cross-checking can be an onerous job requiring one person to scan the manuscript while another ticks off references in the terminal list. Such checking is, however, mandatory unless you are using reference-formatting software such as Endpoint.

TIP

Here is a tip on how do that checking alone, and faster, with a word processor. Print the literature cited section and keep it next to the computer with a pencil. Then search the text using the search/find/goto type of function in your word processor, looking for every occurrence of "19." Of course, that search inevitably brings up the occasional irrelevancy (e.g. 19 April, 19.3 m), but it does find all the citations in the twentieth century. (Well, of course, you should be citing Darwin too, but you can check the few references in the nineteenth century by hand, or make another search using "18.") Tick them off your hard copy literature list as you find them.

Probably the most troublesome citations to check are references by the same author in the same year (e.g. 1990a and 1990b). In most cases, the order in which such references appear in the terminal list is arbitrary, and so a common practice is to designate the first cited in the text as (a), the next as (b), and so on. If you revise the text, however, the original order of citation might become scrambled, so give this problem special attention when checking references.

TABLES AND FIGURES

Tables, figures, and figure legends are special. For a journal report, these items are placed at the end of the manuscript and then integrated into its body by the typesetter. Reports that will be duplicated from the original manuscript, as well as some theses and other student products, may permit or require integration of tables and figures in the body of the report. Journal publication is now done almost entirely electronically, and most journals ask for tables and figures in electronic form.

Tables

As indicated by its very name, a table consists of tabulated material in columns. A table is an entity that can be typed; if it cannot be typed, then it must go as a figure. Each table should be submitted on a separate page. Formerly, tables were numbered by capital Roman numerals, but most journals now use Arabic numerals. Tables are collected near the end of the report because each is (or, at least, historically was) typeset separately and then inserted at the appropriate place in the text by the typesetter. Tables inserted into a word-processed file are now common. Student reports and theses, however, commonly place a table on the page following its first mention in the text. Wherever placed, each table should be on its own separate page.

Tables generally consist of a title, top horizontal ruling, column heads, one internal horizontal ruling, lines of entry, and bottom horizontal ruling. Many journals allow tabular material to be footnoted; if so, the footnotes go below the bottom horizontal ruling. The title of a table should be informative, stating what the table contains.

Many journals allow explanatory sentences to follow the title. Example from the literature:

> TABLE 1. Durations, highest frequencies, and lower frequencies of note types in chick-a-dee calls of the Mexican Chickadee. Entries are mean ± standard deviation (sample size).

Figure legends

Figure legends are typed together, beginning on a new page, following the tables. This practice arises from the fact that typesetters originally set the type separately on figure legends, and then inserted them under figures within the main body of the paper. On student reports and theses, however, it is usual to place the legend beneath its figure on the same page. Also, see the section *Types of figure* for notes on plates, which may be numbered separately from text figures.

If a figure has several parts, label them with lower-case letters (unless the journal has alternative instructions). In order to explain all these parts, the figure legend is often noticeably longer than the title lines of a table. Example from the literature:

> FIGURE 1. Sound spectrograms illustrating the A and D notes of chick-a-dee calls of the Mexican Chickadee, selected to illustrate some of the phonological variation found in notes. The frequency and timescales are the same for all spectra. (a) Commonest of all calls is the call type AD, with typical down-slurred A note. Notice the banded structure of the D note. (b) A call showing the typical A/D contraction of the final A and first D notes of calls containing both types. (c) The contracted A/D may begin a call; notice the noisy (unbanded) structure of the D notes. (d) Rare phonological structures occur, as in this call with an A-like introductory note followed by an A/D-like structure of long A component and short D component.

The title of a figure legend is a title (not a grammatical sentence), as in the case of title lines of tables. The "title" of each part may also be written as a title. Anything else in the legend that follows such titles is written in complete sentences.

Figures

Figures are anything that cannot be typed, and they should be included at the very end of the manuscript. Figures are any kind of illustrative material: drawings, graphs, diagrams, and so on. They are all called figures and are numbered serially without distinction by Arabic numerals. Figures are usually placed on the same size of paper as the rest of the report, although some journals allow submission of original figures of any size. Each figure is on a separate page, and the lettering on a figure should be *large* so that it can still be read adequately when reduced to journal size. Using a 14-point sans serif font for computer-created figures usually works well for most journal sizes.

TIP

Here is a tip on how to judge the adequacy of lettering on figures. For computer-produced figures, the graphics program often has a way to view the figure at a small size, comparable with the size of figures in the printed journal. If that provision is not available, then it is sometimes possible to print a draft of the figure at reduced size. If even that procedure is not possible, or the figure is not produced on a computer, take the full-size figure and reduce it by photocopying. If reducing the figure to journal size makes the labels difficult to read, increase the size of the lettering on the original figure.

A few journals require unlabeled figures so they can set the labels in type or perform the electronic equivalent. In this case, you must make a labeled version as well for use by the editor. A sans serif font is preferable.

The actual construct of a figure depends upon the nature of the material being illustrated, so it is impossible to delve into details here. Some issues are discussed in Chapter 4 on presenting research. Suffice it to mention here that "three-dimensionalizing" graphs – which is easy to do with computer software – is poor practice. Bar graphs in particular suffer from adding this visual distraction that has no information content. In general, keep figures as simple as is consistent with their content.

Types of figures

By far the commonest type of figure is the *line drawing*. Early line drawings were appropriately named as they were literally drawings composed of lines, usually made in carbon ink. Today, most line drawings are produced by the author or illustrator on a computer and rendered by a laser printer for submitting to a journal or submitted electronically. Shading in a line drawing is achieved by stippling or hatching or by using a shading tool in a graphics package.

Drawings that use shades of gray (e.g. pencil and charcoal drawings), as well as any photographs and computer-produced illustrations using a true gray scale, are known as *halftones*. "Photographs" in newspapers are halftone illustrations made from true photographs. The finer the grain of a halftone illustration, the more expensive it is to produce.

Journals vary with regard to the production of figures. Most journals require final figures in encapsulated postscript (EPS) or tagged image file (TIF) format, although other formats may be acceptable for initial submission for review. For many journals, all final figures must be in a minimum resolution of 600 dots per inch (dpi). Color figures need to be in the CMYK (cyan, magenta, yellow, black) standard, which translates to printing inks. Figures in the RBG (red, blue, green) screen standard can be converted to CMYK by image-handling software. CYMK is more restrictive than RBG in the number of color shades that can be rendered, and so converted images may need to be adjusted. It is imperative to follow the instructions to authors for the specific journal to which the manuscripts is submitted.

The paper used in many journals was once not of sufficient quality to support fine-grain halftones, and so a piece of special (glossy) paper was inserted for such purposes and designated a *plate*. Although plates are rarely required these days for black and white figures, plates may still be used for color printing, which is very expensive. The physical placement of plates varies widely among journals: each plate may be inserted close to where it is cited in the text; all plates for a given paper may be placed together at the end of that paper; all plates for an entire issue may be collected in the middle of the issue; or all plates may be collected at the end of the issue.

Some journals number figures continuously, regardless of whether they are reproduced on the text pages or on specially inserted pages, but most journals number plates separately from text figures. Color plates are almost never accepted for publication without financial support for their production and printing – either by contribution from the author or from special funds that some societies or journals have specifically for this purpose. In journals where color printing requires separate plates, color figures should be used only where color is essential to understanding the figure fully. Some journals now print color figures within the text and require only that color be useful, not critical, in understanding the figure.

Much of the foregoing is no longer of concern in the context of modern technologies, but we include it for reference. Most journals of developed countries now use paper sufficient for printing gray scale including "black and white" photographs. Reproduction by photo-offset is easier and more forgiving than setting of type, and some journals now use a printing technology that allows color overlays on figures. Basically, anything you can now print on standard-size paper in a laser printer is acceptable to journals using modern technology. The situation with color, however, continues to require the checking of recent issues of a journal and its notes for authors. Formerly, many ornithological journals had a frontispiece, a color plate as the first page of each issue. These were painting or color photographs of birds. The *Wilson Bulletin* still does this, but the *Auk* prints such former frontispieces on the journal's front cover.

How to cite tables and figures

As tables and figures are created separately and then inserted into a paper by the typesetter, the manuscript must indicate where each is to be placed. There are two primary ways to so indicate, some journals preferring one, some the other (a few do not care which). The first is to type into the manuscript a centered (or sometimes left-aligned) note such as:

Table 2 about here

This system has the merit of being embedded in the manuscript and able to be moved around in a word processor if the paragraph referring to the table is moved by cutting and pasting. The other method is to make a marginal pencil note of the same content. The disadvantage of this system is that every time you revise, you must go through the manuscript and pencil in all those notes all over again. Besides the unnecessary labor, that is the kind of repetitive procedure that leads to mistakes, and it is not feasible when your submission is by electronic means.

In general, notes on the placement of tables and figures should appear after their first reference in the text. We like to place the note at the end of the sentence or, ideally, the end of the paragraph in which the table or figure is first cited; putting it within the sentence or paragraph distracts editors and reviewers. It is permissible to refer ahead to a table or figure that really belongs later, but this sort of procedure should be used only when necessary. The statement might be something such as: ". . . although exceptions do occur (see Fig. 4, below)." In this case, the figure is not actually inserted at the first mention in the text, but instead occurs later as indicated by the "below."

Tables are almost always cited as "Table 2," but figure citations may be abbreviated ("Fig. 5" in some journals, "Figure 5" in others).

Most journals use initial caps in such citations but others do not. A few journals require that the citation be in full caps: TABLE 2 and FIG. 5 or FIGURE 5.

This section mainly concerns publishing research reports: choosing a journal, writing a covering letter, and events after submission. Nevertheless, many of the principles also apply to graduate or under-graduate theses and to other kinds of research report.

Friendly pre-review
It is a good idea, and indeed standard practice among most of us in science, to prevail upon at least one colleague for a "pre-review" of the finished manuscript. What you consider to be a complete and polished manuscript may not strike others as such, and it is better to discover problems before submission so that appropriate revisions can be made. Welcome, and indeed expect, the opportunity to return the favor in the future. In general, the better you know the people asked, and the greater their scientific competence in the subject area of the manuscript, the better the feedback. Those who are personal friends as well as professional colleagues are most likely to be straightforward in their valuable criticisms. One of us asked a former postdoctoral associate and good friend to pre-review a manuscript, and he said bluntly that the entire study needed to be redone. Three years were required to follow his advice, but the solid publication that resulted was well worth the delay.

Where to submit
Choosing the journal to which the research manuscript will be sub-mitted often requires some explicit consideration. Many studies in behavioral ecology, for example, are potentially appropriate to two or more entire genres of journals. A study of feeding and foraging of monkeys, for example, might be appropriate for journals devoted to primatology, or behavior, or ecology. More general journals are also

often appropriate. Specific studies that would nonetheless be of interest to a wide readership might be submitted to journals such as *Science* or *Nature*. If you have your manuscript pre-reviewed (foregoing section), ask the friendly reviewer for his or her suggestions of appropriate journals.

Deciding where best to submit could depend upon many factors, such as the following. The first consideration is usually to reach the targeted readership. If the main new finding in a study of monkey foraging involves a general behavioral principle, for example, then a behavior journal may be somewhat more appropriate than those devoted to primates or others focusing on ecology. Another consideration is where closely related literature has appeared in the past. The readerships of such journals are already primed, having read many of the papers by others to which your manuscript refers. Somewhat opposite to following the thread of contributions to a given journal is the desirability of spreading over different readerships separate but related papers on a given topic. This consideration frequently applies to publishing papers based on a doctoral dissertation, where you want to alert workers in several different fields to your research. For example, three different chapters of a thesis on birdsong might go variously to an ornithological journal, a behavioral journal, and a journal devoted to bioacoustics. Yet another, more subjective, consideration is the prestige of the journal. Researchers often perceive a journal as particularly good (or particularly bad), although these reputations can change over time with changes in editors, new directions in research, and more or less random factors. Authors often send their manuscripts to the ''best'' journal they think might publish them.

There are also other kinds of considerations that might be relevant to your choice of journals for submission. Many journals print the submission and acceptance dates of papers published, so by checking a recent issue you can judge whether the processing time is attractively quick, reasonable, or inexplicably slow. Journal format might also be a consideration. For instance, if your manuscript has one or more large tables, you would not want to send it to a journal that forbids large tables or formats them badly.

Whatever your final choice of journal, we emphasize again the importance of following its instructions for authors meticulously before printing the final manuscript for submission. Furthermore, you may wish to go back and tailor the introduction to the specific readership of the chosen journal. The more general the journal, the more general background you must provide for the potential readers.

The covering letter

Manuscripts submitted to a journal for publication must be accompanied by a covering letter. The very first thing to determine is to whom the letter with manuscript should be sent; look for directions on the inside cover of the latest issue of the journal or in the journal's instructions to authors online if electronic submission is required. In most cases, all manuscripts go to one specific person or place: the editor (or editor-in-chief) or an editorial office. There are exceptions, however. For example, *Animal Behaviour* – published jointly by the Association for the Study of Animal Behaviour in the UK and the Animal Behavior Society in the USA – has separate executive editors for manuscripts from the Old and New Worlds. Another journal is divided into sections, each of which has an associate editor to which manuscripts for that section need to be directed. Yet another journal directs that manuscripts may be submitted to any person listed on the editorial board.

The covering letter should be formal and businesslike. Senior researchers often know personally the editors of journals in their fields, and so use first names in the salutations ("Dear Mary"), but leave personal comments to separate private correspondence. Usually the greeting will be "Dear Dr X" or "Dear Editor Y."

The opening line of the letter should get straight to the point, stating the title of the enclosed manuscript, its authorship in sequence, and the fact that it is being submitted for publication in the named journal. Although this content appears self-evident, without it misunderstandings can arise. One of us, while editor of a journal, received a manuscript with a covering letter that was vague and was obliged to write back for clarification. It turned out that the manuscript was not being offered for publication but was being sent merely to elicit collegial comments. Here is a sample opening sentence: "Please find enclosed an original and two copies of the manuscript entitled 'Evidence for natural selection in earthworms' by T. H. Huxley and C. R. Darwin, which is submitted for publication in the *British Journal of Evolution*." Online submissions might begin: "Attached please find . . ."

Many journals specify further required content of the covering letter, so it is important to check the journal's instructions to authors. Some or all of the following may be mandatory: (1) identification of the corresponding author for the manuscript (usually the author writing the covering letter for submission) and how to contact this person (postal address, email address, fax number, and telephone number); (2) a statement that all other authors listed on the manuscript have agreed to its final form and to the submission; (3) a statement that the material does not overlap extensively other manuscripts and publications and that the manuscript submitted is not under review in another journal; and (4) a statement that applicable laws, regulations, and guidelines for the use of organisms were adhered to in the research reported. This last point embraces animal care for studies of whole animals, pathogens for disease-related research, DNA procedures for recombinant studies, and so on as appropriate.

Some journals invite the authors' suggestions of appropriate reviewers, although editors are never obliged to follow the suggestions. If you suggest reviewers, be certain to include contact data when you know them or can find them out: postal address, email address, fax number, and telephone number. Suggested reviewers should not have seen the manuscript previously or know much about the specifics of your study; they need to be in the position of mimicking any reader of the journal picking up a new issue and seeing the papers for the first time. Furthermore, suggested reviewers should have no conflict of interest or close ties with the authors that would bias their reviews. Many editors are also receptive to comments concerning people you believe would be inappropriate reviewers, with a clear and objective explanation of why they would not be appropriate. The reason offered must go beyond mere differing views on some scientific issue; indeed, objective criticism from those of different outlooks on how to interpret some complex topic or dataset can be the most valuable feedback received. If someone has been personally hostile to one of the authors, or has had a close personal or professional relationship that would compromise objectivity in reviewing, then it is appropriate to explain the circumstances to the editor. The editor may send the manuscript to such people anyway but then judge their criticisms in light of your explanation questioning their potential objectivity.

Be certain to include or attach all enclosures specified by the journal's instructions to authors. In the case of hard-copy submissions, the principal specification is the number of copies of the manuscript. In some cases, an additional set of original figures must be included, which will be retained in the editorial office. Some journals require submission of the manuscript on computer disk, in a specified word-processing format. The disk may be required on initial submission, but in most cases it is provided later after the manuscript has been reviewed and accepted for publication and after corrections have been made. Floppy disks may not hold an entire manuscript with figures. In that case, you will probably have to burn a compact disk (CD), which means you must have access to suitable hardware.

Many journals have shifted entirely to electronic submissions and provide specific instructions for collating the entire manuscript, including tables and figures, into a portable document format (pdf) file. In order to make a pdf document, you must have Adobe Acrobat software, sometimes now shipped with new computers, or a word-processing program that permits you to do so. In some versions of Word, for example, you have the option of creating a pdf document under the print panel. In making a pdf file, you may need to paste figures drawn in graphics or other software programs on to pages at the end of your text and then save the entire document in pdf, which can be downloaded directly at the time of submission. Also, be aware that electronic submissions via the World Wide Web may entail more problems than email does. You will need to have suitable software browsers (e.g. Netscape™, Explorer™), and in some cases older versions are not fully compatible with interactive aspects of websites. Do not hesitate to contact the editorial office by phone or email if you experience difficulties with the journal's website.

The review process

As with most grant proposals (Chapter 2), the review of a manuscript submission is usually a multistep process. You should receive notification that your manuscript has been received. If you have not heard anything within a reasonable time after submission by mail, you should contact the editor. Confirmation that an electronically submitted manuscript has been received is often immediate, so be alert for email messages confirming that your submission was transmitted. If no such confirmation is forthcoming, you may wish to contact the editor to determine whether your submission was received before you try to send it again.

Journals vary in how much autonomy the editor or editorial assistants have in deciding about your manuscript. Some of the large general science journals, such as *Science* and *Nature*, screen submissions before deciding whether to send them out for review and return a majority of submissions without reviewing them. Most specialized journals will send your manuscript to reviewers for evaluation soon after receiving it. Many editors now make an initial inquiry about the interest and availability of potential reviewers, which permits them to seek alternative reviewers more rapidly if necessary.

As with grant submissions (Chapter 2), there is a fine line between nagging an editor too early in the review process and asserting your legitimate rights as an author to know the status of your submission. Granting agencies usually specify in their application materials how soon you can expect to hear from them, but journals rarely make such specific commitments. Many journals indicate at the beginning or end of a published article the dates that the manuscript was received and accepted for publication. Some also indicate the dates of resubmissions. You can use these dates to gauge roughly how long each step of the review and decision-making process may take, but it is rarely possible to distinguish between an extended review process and an author's delay in supplying revisions. In general, we recommend that you wait at least three months after receiving notification that your manuscript has been received before querying the editor about its status.

Many journals track submissions electronically. Typically, the corresponding author is given a tracking number and can query on the website or by email for information about progress in processing. The returned information might be, for example, that your manuscript was sent to two reviewers on such-and-such a date and another reviewer soon after, and that one reviewer has already replied.

When the editor has received reviews of your manuscript, he or she will usually add his or her own evaluation to a letter summarizing the reviews and the publication decision. Reviewers will usually be expected to recommend whether your manuscript is publishable as is or with minor revisions, publishable with major revisions, or not acceptable for the journal to which it has been submitted. In many cases, these summary recommendations are for the editor's eyes only and not sent to the author. If the reviews are consistent, the editor will usually follow their recommendations. If the reviews vary, the editor's judgment of the merits of any criticism will come into play. The editor may, at his or her discretion, send the manuscript to one or more additional reviewers for further advice before making the final editorial decision.

Upon receiving the editorial decision
Try to be emotionally prepared for a rejection and for an editorial non-decision that requires extensive revision before acceptability for

publication can be assessed. Even experienced researchers can be stunned by criticisms of a manuscript they worked on so very hard for so very long. One of our colleagues advises looking over the editor's letter and comments, and then setting aside the entire package for a week or two before doing anything further with it.

After your emotions have stabilized, you may find that things are not as bad as you initially thought. An outright rejection might turn out to be based on the subject matter being clearly more appropriate for a different journal, with laudatory comments on the manuscript and useful suggestions for a more appropriate outlet. Or, revisions requested may turn out to be less overwhelming than they seemed during the initial shock. One researcher told us that when a manuscript of his is rejected on grounds that he views as inadequate, he sends it to a higher-quality journal, where it usually gets a better review.

Even a manuscript that has been accepted for publication is likely to have at least minor suggestions for revision. One of us had a manuscript accepted for publication after an unusually long (12 months) review process, with the sole request to add a figure. Usually, however, the critiques of a manuscript will be more substantial, and you will need to decide whether you can incorporate the suggestions into a revised version of your manuscript or whether you need to restructure and rewrite the manuscript entirely. You may even conclude that it would be more efficient to resubmit to another journal – but remember that this will entail reformatting the manuscript appropriately and subjecting it to another complete review process. Furthermore, you would need to inform the first editor that you are withdrawing the manuscript from further consideration for that journal.

If you are able to revise your manuscript, then there are steps you can take to reduce the probability that the editor will send your resubmission out for a second external review. Include in your cover letter a detailed, itemized explanation of how you have addressed specific concerns and modified your manuscript in response to reviewers' comments. You might indicate, for example, that you have followed the recommendation of reviewer #1 to include a paragraph clarifying your sampling techniques in the methods section. If you disagree with a reviewer's suggestion, be sure to provide a strong justification for why you did not make the recommended change.

The sooner you return your manuscript with the requested changes, the sooner it can move towards possible acceptance and publication. Depending on the journal and the magnitude of the revisions, an editor may be satisfied that you have met reviewers' concerns and accept the revised manuscript for publication, or the editor may send the revision and your explanation of changes to some or all of the people who reviewed the original submission. In the latter case, the review cycle begins anew.

Once you have received notification that your manuscript has been accepted for publication, it is appropriate to refer to it as "in press" in other manuscripts and on your c.v. (see Chapter 5). Submitting a manuscript does not guarantee that it will be published, but until you submit it most definitely will not be.

How to review a manuscript

Good preparation for writing a manuscript that will survive journal review and for revising a manuscript that has been reviewed is to review a manuscript yourself. A graduate student's first review is likely to be a pre-review of a manuscript by the major professor or a postdoctoral researcher or another graduate student in the same laboratory. Soon after your own papers are being published, editors will probably begin sending you manuscripts on related topics.

A common procedure is to email you to ask you to look at a manuscript (or, more commonly, its abstract) on the journal's website and inform the editor if you will be willing to review it. Or, the editor may send the abstract to you by email. In either case, in order to review the full manuscript you will have to download it from the website. In almost all cases, the editor will provide you with the necessary password, and generally that will give you access only to the manuscript you agreed to review. You may also have to download the journal's instructions for reviewing, and very likely you will have to submit your review electronically as well.

Editors usually provide reviewers with specific instructions, which range from a paragraph of guidelines and suggestions to a rating sheet of specific points. As an example, here are the evaluation items from the *Journal of Comparative Psychology*, provided by our colleague Charles T. Snowdon. Each item is to be rated as high, medium, or low, with supporting comments to be provided on a separate sheet:

- Significance of the topic.
- Appropriateness for this journal.
- Quality of research (design and analysis).
- Quality of writing (organization, clarity, style).
- Ranking with respect to published research in comparative psychology and animal behavior.

In evaluating these items, the reviewer should be as specific as possible. For example, under "Significance," state what is new or important in this contribution (even if this merely restates what the manuscript already says), or explain why you think it is not as new or important as the manuscript claims. Quality of research is very important: Do the conclusions drawn follow inevitably from the results? Are the experiments conducted logically and capable of rejecting the hypothesis being tested? Were controls adequate, and was statistical analysis appropriate? If you recommend that something different should be done, be as specific and helpful to the author as possible.

The letter from which this list of evaluation items was taken also asks two other specific questions of reviewers. First: What other journals might be more appropriate? A legitimate answer to this might be "None." Second: Do you have any ethical concerns with this manuscript? This question could cover a variety of ills, such as treatment of research animals or human subjects, safeguards in DNA and disease studies, plagiarism, and so on. The reviewers' responses serve to alert the editor to a potential problem that he or she will pursue. Also, see Appendix B on ethics considerations.

Finally, the letter requests from the reviewer a recommendation from the following list of possibilities. This list is more extensive than that used by most journals, so it shows in detail the range of conclusions that a reviewer might draw with respect to a particular manuscript:

1. Strongly recommend acceptance:
   (a) as is;
   (b) with slight revisions as noted.
2. Recommend acceptance with some reservation:
   (a) paper worth publishing but not exceptionally important or substantial;
   (b) needs a few major revisions, as described in my review.
3. Potentially publishable manuscript. Needs extensive revision (rewriting, reanalysis, etc., but could be made publishable).
4. Doubtful. Probably should be rejected. Needs major revisions and even if undertaken would still probably be borderline.
5. Reject (please indicate major reason):
   (a) contribution not substantial enough;
   (b) not appropriate for this journal;
   (c) major methodological flaws that preclude publication;
   (d) other.

The editor will almost always send a copy of your written comments to the author. In many cases, your completed rating sheet will also be included. Some editors, however, do not want the author to see specific recommendations concerning suitability for publication because these might conflict with the final editorial decision. Therefore, unless instructed to the contrary, do not include in your written comments such a specific recommendation; provide that only in the proper place on the reviewer's form or in a covering letter to the editor.

Above all, be objective and helpful in a critique. Keep in mind that you are reviewing a manuscript, not a person, so *ad hominem* comments have no place in your review. After drafting your review, read it back, trying to put yourself in the place of the author who will receive it. Were you fair? Were your criticisms specific enough? Did you phrase things so as to penetrate the mind rather than stir the emotions?

Journals' policies vary with regard to identification of reviewers. Some require that reviewers remain unidentified to the author under the assumption that anonymous reviews will be more frank. Others strongly urge referees to sign their reviews under the assumption that signed reviews will be more objective. In fact, a few journals require reviewers to identify themselves. Most journals remain neutral, stating that if the review is signed at the bottom, then the author will see it, and if not, then the reviewer will remain anonymous.

It is difficult for us to offer advice on whether to sign a review. One of us has signed every journal review right from the beginning of the career, and in a few cases this policy has led to personal animosity from an author. In many more cases, though, authors have specifically expressed appreciation for the identification. Most scientists probably never sign any journal reviews. A few follow a procedure that strikes us as inconsistent and perhaps not completely honest: these few sign laudatory reviews but not critical ones, thus attempting to make friends while avoiding making enemies. So all the models are out there among senior scientists, and each new reviewer coming into the arena must make his or her own choice as to which to emulate.

A manuscript received for review is the intellectual property of the author, and in the USA is protected fully by copyright laws from the moment of its creation. Therefore, manuscripts should be treated as completely confidential material. Upon occasion, a senior reviewer may wish to have comments from a graduate student or postdoctoral associate in his or her laboratory to enhance points in the review. This procedure is usually acceptable, provided that the manuscript is not photocopied and the auxiliary reviewer is specifically named to the editor so that there is a record of who has seen the manuscript.

Passing the manuscript outside of the reviewer's laboratory for additional comment by others should not be done without explicit permission of the editor in advance.

Many journal editors and program officers of funding agencies ask reviewers to identify any potential conflicts of interest they may have with the author(s) of the manuscript. Conflicts of interest might include an unusually close relationship, such as that between an advisor and student, relatives, or friends, or an antagonistic relationship, such as that between direct competitors in a small research area. Any time you are uncertain about your ability to review a manuscript or a grant objectively, you should alert the journal editor or program officer to the basis of your concern. He or she may still ask you to provide a review, but at least they will be alerted to any unintentional biases you may hold.

Finally, what to do with the manuscript when your review is complete? Many journals encourage reviewers to mark directly on the manuscript and return it with the review. Other journals want all comments to be supplied separately in writing, so that the editor and the author can each have copies. In such cases, reviewers are usually instructed to destroy the manuscript, which they should do even if not instructed. Under no circumstances should a manuscript received for review be retained, copied, or even shown to others without explicit permission of the author and editor. Manuscripts received for review should never be quoted, and results or conclusions should never be used by a reviewer before the paper is published. We do recommend, however, that you keep a portfolio of the reviews you write, as the number of different manuscripts reviewed for different journals could be a useful line in your c.v. (see Chapter 5).

Before submitting a manuscript of your own, you may find it helpful to try reviewing it as if you were reading someone else's submission. Reviewing your own manuscripts may enable you to evaluate them as your reviewers will, and to catch and fix any problems in advance.

4 How to present research

The speaking in perpetual hyperbole is comely in nothing but love.

Francis Bacon (1561–1626)

The science that is discussed in a scientific presentation, like the science described in written proposals and reports, can fail to make an impression if the presentation is lacking in clarity, organization, and interest value. Therefore, learning how to give effective scientific presentations is as important in communicating your research findings as any written product.

The three main types of scientific presentation are seminars or colloquia, oral papers at professional meetings, and posters. Seminars and meeting papers are spoken presentations that differ primarily in length and, consequently, content; poster presentations are mounted on a panel and read by passersby. In this chapter, we provide suggestions for how to prepare and present scientific seminars and both oral and poster papers. Most of our advice is derived from our own experiences as speakers, members of audiences, and session or seminar organizers. In addition, Appendix A provides tips on how to write clearly, many of which apply to construction of slides for oral presentations and creation of poster papers.

Seminars are usually allotted an hour or so, with the expectation that the speaker will talk for about 45 minutes and will answer audience questions during the remainder of the time. Meeting presentations are usually limited to 15 or 20 minutes each, with the expectation that the speaker will talk for about 12 or 17 minutes, respectively, and respond to questions during the remaining minutes. Limits on presentation time at meetings are usually enforced strictly in order to permit members of the audience to move freely between concurrently scheduled talks, but even the more flexible seminar hour should be respected because members of your audience may have to leave for other appointments.

Scientific presentations, even more than written reports, reflect each researcher's unique style and personality, and what works for some individuals may appear forced, and therefore ineffective, for others. Evaluating effective presentation techniques at seminars and other talks is as educational as attending to their scientific content.

RESEARCH SEMINARS

The most common type of hour-long research presentation is the departmental colloquium in a large research institution. In this context the audience comprises largely faculty, postdoctoral associates, and graduate students in your own discipline, but few will be experts in your specialty. In smaller institutions, the entire division of sciences, or even the whole college, may host a seminar series, so the audience is even more diverse. It is crucial for a doctoral student to develop a polished hour-long presentation of his or her research because the departmental colloquium is a standard part of the recruitment procedure for hiring, affectionately known as the "job seminar."

The general plan of a research seminar is straightforward: introduce your study, explain the findings, and summarize the study and its implications. Robert G. Jaeger, a distinguished expert on the behavioral ecology of salamanders, is fond of quoting a mythical southern preacher on how to give a sermon: "First ya tells 'em whatcha gonna tell 'em, then ya tells 'em, and then ya tells 'em whatcha told 'em." Do not shy from redundancy: even major points are sometimes missed by attentive listeners during oral presentations.

Advance material

When you sign up or are contacted to give a research seminar, you should be prepared to provide a brief, informative title for your talk, a paragraph-length description of your talk, and a short autobiography indicating your name, your current status (e.g. doctoral candidate, Ph. D., postdoctoral fellow), and your institutional affiliation. If posters to advertise your talk will be distributed, then you may also be asked to provide a line drawing or photo to illustrate the announcement.

You may need to know a number of things from your host in order to plan effectively. If you are unfamiliar with the seminar or colloquium series, you will want to confirm the length of time you will have to speak and answer questions. You may also want to inquire about the composition of the audience that usually attends the series. Knowing how many people typically attend, and whether they are a mixture of students and faculty, mainly faculty and post-docs in a specialized area, or an interdisciplinary group will help you to plan the level of your presentation.

You should request in advance any special equipment you may need for your presentation. If the room is large and you have a soft voice, then you may want to ask about access to a microphone. If offered a microphone as a matter of course, it is wise to accept it because your host probably knows the acoustics of the lecture hall better than you can judge. We have attended talks where the speaker refused a microphone as unnecessary, and then we strained to catch some of what he had to say. Even senior researchers have made the mistake of refusing a microphone when offered by the host.

Other items of request might include a pointer to help highlight key data on the screen, a podium lamp or small flashlight if you want the option of being able to refer to your notes while showing images in the dark, and a blackboard or dry eraser board if you want to be able to draw while you speak. You should also alert your hosts to your computer needs. If you bring your own computer, be sure that you also have your power cord and any necessary attachments to connect your computer to the projector provided by your hosts. If you bring your presentation on a storage device, such as a CD or USB stick (flash drive), be sure to confirm in advance that the computer you will be provided will be able to download your presentation. It is usually possible to advance your own slides directly from the computer or with a separate remote control, but it is worth confirming this because the layouts of lecture and seminar rooms can vary quite a bit. Knowing whether you will be able to control your slides yourself or will need to request someone else to advance them for you will help you to gauge the length of your presentation. We will return to the logistics of preparing computerized presentations after considering their content.

Content and organization

Know your audience and prepare appropriately. If a colloquium attracts faculty, postdocs, and graduate students from diverse disciplines, then you can assume that most of your audience will need you to establish the scientific context and research questions that your study addresses before you delve into your results and conclusions. In this respect, a seminar presentation resembles a research report (Chapter 3) and includes an introduction, review of methods, results, and discussion.

Seminar presentations can, however, be structured more loosely than research reports. Use of the first person throughout is appropriate and helps to establish a more personal level of communication with your audience. Interspersing a few relevant anecdotes will contribute to a more interesting and engaging presentation. Think of your presentation as an opportunity to tell a story about your research. You want your audience to pay attention, to understand, and to try to anticipate what will come next. To help your audience to accompany you in these ways, you will need to provide the necessary background to the research, your study subjects, and your observational or experimental protocol. The most critical parts, of course, are your data and the conclusions you draw from them, but if your audience has lost track of why these data are important or lost interest in hearing about them, recapturing their attention may be difficult.

Organize your talk by developing an outline from the results section backward to your introduction. Decide what data you want to discuss, and work back to the methods that you will need to describe and the introductory material you will need to cover to set the stage. Then work forwards to your discussion, being sure that you cover every question you raise in your introduction. You may also find it helpful to use your outline to decide how much time you will devote to each part of your talk.

Timing is very important. Forty-five minutes may seem like a fairly long time, but it will pass rapidly once you begin to speak. Both of us have attended poorly organized seminars in which the speaker is just beginning to describe his or her results when the hour ends. A well-organized seminar ends promptly, repaying the audience for its attention and allowing respectful time for its participation by questions. Roughly, your seminar should be divided into a five- to ten-minute introduction to the problem, a five- to ten-minute discussion of methods (including study site and study subjects), a 15-minute demonstration of your data, and a ten-minute discussion. Especially if your audience is diverse, you may want to spend longer at the start in laying out the nature and importance of the research, making up the time by briefer explanation of the methods and shorter terminal discussion. Planning 40 minutes of substance allows time for the host to introduce you and make any routine announcements and for a few brief digressions into anecdotes or repetitions if you sense that your audience needs further stimulation or clarification. We recommend that you prepare a few anecdotes in advance, but be sure to keep them brief so that they don't cause you to run out of time.

Practice

Giving a practice seminar to colleagues and friends will enable you to weed out unnecessary slides or choppy passages and to identify where a transitional "filler" slide might help you along. Practicing will also give you confidence in public speaking and allow you to work out your timing and any confusing areas in advance. Practicing a seminar talk aloud, like revisions on written proposals and reports, is essential for a fine-tuned product. Even experienced speakers who have developed a good sense of how long it will take them to explain each of their points often practice their talks out loud while looking at their slides. As the rule of thumb for writing is "revise, revise, revise," that for speaking is *practice, practice, practice*!

Style and delivery

If possible, arrive early to survey the room in which your talk will be given. You may connect your computer to the projector before or after the projector is turned on, but do not turn on the computer on until the projector is running. The projector needs to be running so that the computer can recognize it upon startup. If your talk is saved on a storage device, copy the talk into the computer provided for you. Test the connections to be sure that the projector can identify the computer and will project your presentation when you are ready to begin. Ensure that you know how to work any controls that will be your responsibility (e.g. dimming lights, advancing slides). Become familiar with the pointer equipment you will use, such as flashlight or laser pointers. If you have a preference for placement of the podium relative to the projection screen, ask if you can choose the position. Many people like to have the podium on stage right if they are left-handed or stage left if they are right-handed, so that the preferred hand is towards the screen for pointing. One of us, however, prefers the reverse, partly in order to hold a laser pointer more steadily with both hands in front of the body.

It is now standard practice in some fields for speakers to distribute at the outset of their seminars a handout with copies of their slides and space for their listeners to take notes. Ask your hosts in advance if you are uncertain about whether this custom applies in your case, and be sure to come prepared with enough copies if it does.

Your talk begins visually before you utter a word. It is wise to dress comfortably and professionally, some speakers recommending dress that is one step more formal than the expected dress of your audience, in order to show your respect for the occasion and the listeners. Be prepared to stand throughout your talk, even if it is an hour-long seminar, except under extraordinary circumstances. If you suffer from lower-back problems, arrange for a small footstool or other support to raise one foot (there is a reason for those foot rails in bars). Strolling occasionally, but not distractingly, while you talk also lessens the physical stress of standing. One of us gave a departmental seminar after the sudden onset of a viral infection the previous night, and with a temperature of over 100 °F was forced to speak much of the hour from a chair kindly provided by the host. That is what we mean by "extraordinary circumstances" – something that might happen only once or twice in a career. Equally effective speakers may have markedly different delivery styles, and it is important to develop a style that suits your personality and therefore comes naturally, however nervous you may be. Some speakers are comfortable with a conversational, almost chatty style in which they allow the audience to appreciate their sense of humor without detracting from the science they are discussing. Other speakers appear forced and awkward when they try to tell jokes and would be better off sticking to a more formal, straightforward presentation. Whatever your style, your first utterance should be of thanks to the person who introduced you.

If you have practiced giving your seminar with your slides, you may discover that your slides provide enough prompts about what you intend to say that you have no need to consult written notes or text. The more familiar you are with the sequence of your slides, the more easily you can rely on them to propel you along. We caution against actually reading a prepared text aloud. Written sentences are meant to be read and are often too complex to be spoken effectively. Reading also tends to be lifeless compared with naturally spoken words. Listening to someone read a 45-minute seminar is rarely as engaging as listening to someone talk. Nevertheless, you may wish to write out your talk initially in order to help you organize your thoughts. A graduate student once told one of us that he wrote out his presentation and put the script in his pocket for confidence, afraid that he would "go blank" during his talk (which never did actually happen).

Practice will help you to discover how you feel most secure, and self-confidence, both in your science and in your speaking style, will make a seminar presentation more enjoyable for you as well as your audience. Observing other seminars and evaluating what you find most stimulating is helpful as you begin to prepare for your own. One of the authors attended multiple seminars in an entirely different – and nearly unintelligible – field so as to focus exclusively on delivery styles without the distraction of also trying to understand the content. Observing other speakers critically will also help you to identify some of the irritating nervous quirks, such as twirling a pointer, turning away from the microphone, or the use of "uhs," that can detract from a talk. Almost everyone has such quirks, even if expressed only very occasionally. A way to discover yours is to have someone videotape one of your talks. As both of us are experienced at giving talks, it came as quite a surprise to one of us recently to hear so many "uhs" upon playing back a videotape of the talk.

Emphasize the important points that you want the audience to understand and remember most. Several forms of emphasis exist, including modulation of your voice and the use of simple hand gestures. An often effective mechanism is pausing briefly before making an important point; sleepy audiences tend to perk up if the patter ceases. Some speakers may appear blunt in saying aloud "This is important," but the ploy is effective.

Presenting a seminar is part showmanship, but even somewhat dull deliveries will be appreciated if the talk is well organized and intellectually stimulating. Conversely, highly entertaining talks without substance or organization will rarely leave a positive impression. Learning how to give exciting and informative seminars is essential when interviewing for academic and research jobs. Good slides tend to contribute to good seminars, and good seminars tend to generate good questions from the audience. Perhaps above all else be enthusiastic about your study; if *you* aren't, no one else will be either.

Ending the talk

Many speakers find it difficult to conclude their talks, so give your ending some advance thought. Do not trail off lamely with some minor point, and do not ask for questions (that being the job of your host). A traditional way to conclude the talk is by thanking the audience for their attention. Nevertheless, one of us was taught in a public-speaking class never to end a talk in that fashion but instead to conclude with a strong ringing point. However you do it, your words and delivery must make it clear that the end is *now*.

A forceful and useful way to end a talk is with the most important "take-home" message of your seminar. If you wish to give polite thanks to your audience, do it before the take-home point (or following the question period, as noted next). For example, you might say something like "Thank you for your attention over the past hour, and I hope that if you learned nothing else of interest this point was clear . . ." followed by your take-home message. One of us still remembers the take-home ending of a departmental seminar by ecologist Paul Colinvaux in the 1960s. He had made a detailed study of cores from the bottom of lakes in Alaska, and from the pollen at dated times in the cores he could reconstruct the climate during the period of the Bering Land Bridge. The final sentence of his talk was this: "When human beings first came to the New World, it was very cold indeed."

The question-and-answer period
After your conclusion, the lights will be turned on (if you have not already requested them), and the host will usually ask the audience for questions. Depending on audience size and formality, either you or your host will choose whose question is taken first. Sometimes, listeners are immediately vying for the floor, but it is more usual that they are mulling over some point before articulating the question they wish to ask. It may seem an eternity before the first question comes forward, but be patient, for the silent time is almost always far shorter than you perceive it to be. Usually a good idea is to repeat the question before answering it. Even though you at the head of the room may be able to hear the question easily, that may not be true of every spot in the audience.

If there are no raised hands in the audience, it is your host's job to ask the first question. However, you cannot depend upon the host to know his or her job, so be prepared with some brief addition to your talk. A confident speaker may say something like: "While you are formulating possible questions in your minds, let me say a few words about one of the points mentioned in passing." Either your host's polite question and your answer, or your little addition, will usually buy sufficient time for members of the audience to step forward with their questions.

Questions are rarely intended antagonistically, even if they appear to, or quite explicitly do, challenge your data or your conclusions. There is no shame in acknowledging that the question raises a viable alternative to the one you advanced, or that you do not have the necessary information to provide a confident response.

Someone in the audience may say something to the effect "This is probably a dumb question, but . . ." We believe that questions are rarely (perhaps never) truly dumb. If someone failed to understand something in your seminar, then that could be due to your presentation. A nice practice is to assure the questioner that dumb questions don't exist. Also, sometimes a question is not well formulated, so you can restate it before answering. Such a question might strike others in the audience as being dumb, but by restating it you can skillfully bring out its pertinence. Always try to deal with questions in ways that will encourage more questions.

The question period can extend your talk by enabling you to discuss tangential points or to speculate about what you may suspect or would predict. If several people indicate that they have questions, you will want to limit the length of each reply so that you have time to take as many questions as possible. If your audience is slow to respond to the call for questions, then you may indulge in your first answer in order to give others more time to formulate further questions. Use the number of hands that are raised at each call for questions to gauge how concise or lengthy your responses should be. Thanking the audience at the end of the question period, indicated either by the time or the lack of further questions, is a positive way to express your appreciation for the opportunity to speak.

AUDIOVISUAL AIDS

Any talk, including departmental seminars just discussed and oral meeting papers discussed in the next major section, profits from good slides. This section tells you how to prepare slides for computerized presentations and use PowerPoint®, discusses playback of recorded sounds, and adds notes on other audiovisual aids.

Preparing slides: fundamentals

Most seminar presentations rely on projected images to illustrate and emphasize the accompanying dialog. In the past, these images were 35-mm slides that were loaded into a carousel and projected from a slide projector. Although most college and university campuses are still equipped for projecting 35-mm slides, it is now widely assumed that speakers will show digitized images in computerized presentations, and many professional meeting sites throughout the world have shifted exclusively to them. PowerPoint presentations have become synonymous with computer presentations owing largely to the widespread availability of the PowerPoint software and its compatibility across different types of operating systems. If you prepare your presentation using another kind of software, you may still need to transfer it into PowerPoint to be compatible with the computer from which your presentation will be projected.

There are three great advantages of computerized presentations over traditional 35-mm slideshows. First, computerized presentations are less expensive to produce, provided one has access to a computer and software, because computer-generated figures and text can be inserted into the presentation directly. This convenience eliminates the intermediate time and costs involved in photographing material from the computer screen or printouts of it, and then developing the slide film. Second, the content of presentations can be changed readily. This advantage saves time and money each time new results necessitate updating data slides and also permits you to fine-tune your presentation up until the last minute. In the old days of 35-mm slide presentations, speakers could switch the sequence of their slides with some ease, but they could not alter the contents of their prepared slides at short notice. Finally, computerized presentations can be animated to emphasize important points as the speaker is making them. Animation settings make it possible to add, delete, or move material within a slide. Other compatible software programs make it possible to insert digitized video clips and sound recordings into the presentation. The result is that computerized presentations tend to be livelier, and require less juggling between equipment, than traditional 35-mm slideshows.

We have found the rapid switch from 35-mm slideshows to computerized presentations to be fairly simple – except for the time involved in digitizing our old 35-mm slide collections and the care that is required to insure color quality and computer compatibility. Many of the same rules about designing presentations and the format and content of 35-mm slides also apply to digitized slides used in computer presentations. In the rest of this section, we review some basic information common to both types of slide preparations, and then discuss some additional factors to keep in mind when developing and presenting images in a PowerPoint format.

Some speakers begin immediately with a slide, often humorous, to introduce their talk; others prefer to establish eye contact with the audience before dimming the lights for their slides. In either case, thoughtful selection and preparation of good slides greatly enhance the ability of your audience to follow and evaluate what you are saying.

There are three basic types of slide: photographs, text, and data. Photographs are often the best way to familiarize your audience with your study animals and their habits, as well as the location of the research. Text slides include stated hypotheses, conclusions, and further questions. You may also have an acknowledgment slide that lists the individuals and institutions that contributed to the research. Present data in either tables or graphs, although graphs are always preferable to tables whenever the information can be displayed with a figure. Entertainment slides, such as pertinent cartoons or artwork, may be included, but too many will detract from the scientific content of your talk.

Text and data slides are only as informative as they are intelligible. Both should be prepared horizontally whenever possible; in computer jargon, horizontal format is called "landscape" orientation. Text slides should be printed in large boldface fonts that will project clearly. Leaving spaces between lines, numbering or bulleting each point, and using indentations where appropriate will help your audience to follow along. Figures and tables should also be printed large enough so that they project clearly. Figure axes should be labeled, and legends and statistical values should be indicated. Tables should include only as many rows and columns as is absolutely necessary; tables with more than five rows and three columns may be difficult for your audience to absorb.

Packing too much on to a slide is counterproductive. Not only do viewers feel rushed in trying to assimilate all the material – sometimes to the point of not listening to what you are saying – but also people sitting at the back of the room may not even be able to read print face that is too small on a crowded slide. The International Congress of Zoology held in Washington, DC, in the 1960s limited slides to eight lines of print, for both text and tabulated material. Speakers had to submit their slides in advance to a central office, where congress officials previewed every slide before sending them to projectionists. Any slide with more than eight lines was simply removed and placed in an envelope for the speaker to pick up after his or her talk. That true story makes for a good rule of thumb:

RULE

Never place on a slide more than eight lines of text or five rows in a table.

The sensible use of animation can enhance a presentation by making the speaker's points easier to follow and the presentation more active. But just as too much information can defeat the point of a text or data slide, the overuse of animations can be a distraction. One of the most common and effective uses of animation is when it is applied to a bulleted list so that each item makes its first appearance on the screen when the speaker is ready to discuss it. Members of the audience can then also read the summarized bullets accompanied by the speaker's elaboration of them. To achieve the same effect with traditional 35-mm slides, one would begin with a slide with the first bullet, and then advance to the second slide in which the second bulleted point was highlighted below the first, and so on.

A sequence of four to five bullets would occupy as many different slides, whereas in PowerPoint one only need to type out a single list and add the animation settings. Be careful, though, because the economy of having fewer slides does not translate into a corresponding economy of time. It will take you as much time to talk through each bulleted item on a single animated slide with four bullets as it would have taken to explain four separate slides with the same information. Consider, for example, a presentation with ten slides, four of which have three sets of data or text that will appear as discussed. This amounts to the equivalent of six single slides plus (4×3) animated slides, or 18 single-slide equivalents. This leads us to another rule of thumb:

RULE

Treat each animation as if it were a separate slide when estimating the length of your presentation.

There are some exceptions to this rule, of course. For example, you may want to talk your audience through the axes and overall data shown on a graph and then use an animated arrow to call attention quickly to a particular point on the graph that leads you into the results shown on your next slide. Similarly, you may begin by showing a digitized photograph of your study subject, and then highlight with an arrow or circle the part of its anatomy, posture, or activity that relates to your next point. In cases such as these, the arrow essentially replaces the pointer that you would otherwise have used to call your audience's attention to the feature of interest. Using animations in these cases can remind you of the emphatic point you intended to make. Keep in mind, though, that too many animations can slow down the rate at which the computer can read your presentation and respond to controls.

Nowadays, most researchers prepare their own text and figure slides on their computers. Text slides can be typed directly into a PowerPoint presentation, whereas graphs may be more easily and more accurately drawn in a graphics program and then either copied or inserted into the PowerPoint presentation on a blank slide. Photographs are also typically copied or inserted from an image file into PowerPoint. While photographs can fill the entire slide screen, be sure to leave wide enough margins on all four sides of any text or figure slides so that edges are not cut off when the slide is projected. Some speakers prefer to use the traditional black print on a white background; others prefer reverse contrast, such as white print on a blue background (traditionally known as blue-line slides). Such reverse-contrast slides are not recommended by professional photographers, because light "leaks" from the bright lines, causing the edges to appear fuzzy when projected. Nevertheless, reverse-contrast slides have become very popular, and many viewers report that they are easier to read in a room that is not very darkened. There are many more color and background pattern options available on most computers, and many combinations produce extremely attractive slides. The software permits you to apply a consistent background over which all slides will appear, thus giving your presentation a coherence that many viewers find appealing. Remember, though, that "glitz" does not compensate for poorly laid out material.

Preparing slides: colors and computers
Although color slides produced on computers are now fairly routine, there are several important considerations to keep in mind while preparing them. It is important to pay special attention to contrast in colored slides. Avoid blue text on a purple background, red on

orange, and similar combinations with little color contrast. Also be aware that five percent of the US male population is red/green color-blind. These people usually have little trouble with traffic lights because of the purposeful redundancies of position (red on top) and special brightness relations of the red and green used, but there are no such additional cues in your slides. Therefore:

> **RULE**
>
> Never use red and green as major color contrast in slides.

Although color increases the attractiveness of slides, more importantly it can be used to help get your message across. For example, major points to be covered can be listed in an introductory slide where the points are color-coded. Then each set of subsequent slides about a given point can use the color of that point to help the audience follow the progression of the talk. Another use of color is to emphasize a special word or point in a special color. Some of the most effective presentations we have seen recently included a judicious mix of color and animation for emphasis. Again, though, we caution not to overdo it.

One advantage of computer-produced slides is the variety of fonts available. Fonts, like colors, can be used to flag points or to emphasize important material. Too many fonts, however, are distracting, and we recommend using the same font type in all slides in a single presentation. We find that varying the size of the font is preferable to mixing the fonts on a slide, or that color or animation is more effective than a different font at calling attention to particular elements of the slide. There are also technical restrictions on fonts used when transferring files between different types of computer and projector, and between different versions of PowerPoint. Consult your hosts or meeting organizers for technical advice before designing slides on your computer, especially if you will not be projecting your presentation directly from it. Finally, studies have shown that serif fonts – those having little marks on the characters, as in this text – promote good reading of text by helping to keep the eye on the line being read. By contrast, sans serif fonts – those lacking such marks, such as used in this aside – are easier to read in orientation other than horizontal and so should be used to label graphics, especially graphics with vertical axes.

TIP

A very effective slide can be made by adding print over a photograph. For example, you can show your study species with its common and scientific names overlaid. Or, you can show an experimental apparatus with its parts labeled. The base photograph can be scanned in from a print with a flatbed scanner or produced from a 35-mm slide using a slide scanner, copied or inserted into PowerPoint, and then written on in order to prepare the final slide.

Photographs in PowerPoint®

Speakers using photographs in PowerPoint presentations should understand the basics of projector resolution and how to prepare and insert pictures. If you use a higher resolution than the projector can display, you unnecessarily inflate the computer memory space required for your presentation. Exceeding the available memory of the computer driving the projector will cause your presentation to fail to load, display slides in a distorted fashion, or crash. If you use your own laptop for your presentation, you will know whether your presentation runs correctly, of course. On the other hand, if you take your presentation on media for use on a different computer, you risk disaster. Photographs that are not prepared properly are almost always the cause of unnecessarily inflated PowerPoint memory requirements.

Liquid crystal display (LCD) projectors that are in common use have super video graphics array (SVGA) or extended graphics array (XGA) resolution, which means full display dimensions of 800×600

and 1024×768 pixels, respectively. Higher-resolution projectors (super extended graphics array, SXGA, and ultra extended graphics array, UXGA) are falling in price and you may come across these at well-heeled venues. A higher-resolution projector will merely show your photograph at its pixel dimensions without distorting it in any way. A lower-resolution projector (e.g. SVGA displaying a 1024×768 photo) will show the photo at the projector's resolution, again without distortion.

RULE

Reduce photograph dimensions to approximately 1024×768 pixels (XGA resolution).

Photographic resolution is a confusing subject because of three interacting variables: pixel dimensions, physical size, and pixel density at a given physical size. Ordinarily, we would not treat a technical subject such as this in such detail here, but the topic of digital resolution is sufficiently important and confusing to merit an extended explanation.

Duplicate your photo, work with the copy in image-processing software (such as the popular Adobe® Photoshop®), and make certain that the width/height ratio will be preserved (usually a checkbox in the software image-size display). Then set the pixel dimensions.

Setting pixel dimensions is easy for a photo taken with any standard digital camera because the picture has the same $4:3$ ratio of width to height as projector displays. Assuming your digital photograph exceeds XGA resolution, just set the width to 1024 pixels and the height will readjust to 768 pixels. Almost all digital photos will exceed XGA resolution, which is about 0.7 megapixels (the product of 1024 and 768). Only the very earliest digital still cameras had a resolution lower than 1 Mp. Nevertheless, when the original digital photograph is of lower resolution, use it as is (i.e. at its maximum resolution) if you really must use it. It is best to avoid using such low-resolution photos because you could display them at too large a size within PowerPoint, where the pixels will show and the photo will appear distractingly "blocky."

Setting pixel dimensions for photographs scanned from slides or prints is a bit more involved because the width/height ratio will rarely be 4:3 (never so in 35-mm slides). Assuming the pixel width exceeds XGA resolution, set the width to 1024 pixels. If the resulting height is less than 768 pixels (as it will be for scans of 35-mm slides), you will not be able to fill the screen with the picture unless you first crop the photo to a 4:3 ratio. If the height that results from setting the width to 1024 pixels exceeds 768 pixels, set it to 768 pixels and the width will be readjusted to some value smaller than 1024 pixels. If the scanned photograph is of lower resolution – width less than 1024 pixels *and* height less than 768 pixels – then use it as is (i.e. at its maximum resolution). If at all possible, however, rescan the original at a higher scanning resolution in order to be able to use the photo at XGA resolution in PowerPoint. If rescanning is not an option, ask yourself whether you can do without the low-resolution photograph.

You can resize the physical dimensions of the photograph after bringing it into PowerPoint by dragging a corner. Such physical resizing has no effect on the pixel resolution of the photo. If your photograph has the pixel dimensions of 1024×768 pixels, you can make it fill the PowerPoint display precisely. If its dimensions are 1024 by less than 768 pixels, you can make it the width of the screen, maximum; if its dimensions are less than 1024 by a full 768 pixels, you can make it the full height of the screen, maximum. If your photo is less than XGA resolution in *both* dimensions, then the maximum physical size is a more complicated issue, which we discuss below. You can also resize the photo in PowerPoint by dragging to any smaller physical size than its maximum.

An annoying inconvenience can arise when bringing a photograph into PowerPoint at the recommended XGA resolution. The photo will often be physically larger than the size of the PowerPoint display. This mismatch happens because the photo is shown on your computer at the greatest pixel resolution possible given your particular screen, while the Powerpoint display is at an automatically reduced size in order to fit your screen. The consequent mismatch means you must find a corner of the photo in order to drag it down to a physical size that fits the PowerPoint display. Because the photo is centered on the PowerPoint display, all its corners may be hidden. Therefore, you must move the photo in some diagonal direction (click and drag inside the photo) in order to find a corner for shift-dragging. Remember, resizing a photo in PowerPoint does not alter its pixel dimensions, so you can always make a photo *smaller* without affecting its projected resolution.

To avoid an annoyingly large photograph when bringing it into PowerPoint, first set the physical size of the image (called "print size" in Photoshop) in your image-processing software. Remember that three interacting factors exist in photo resolution: pixel dimensions, physical size, and pixel density at a given physical size. In Photoshop and similar software we have seen, changing the physical size leaves the pixel density (called "resolution" in Photoshop) unaltered but changes the pixel dimensions. Therefore, you must set the physical size first, and then change the pixel dimensions in order to force the change in pixel density at the new physical size. You should choose some physical size that will make the photo smaller than the PowerPoint display on your screen; that physical size depends upon the size and resolution of your screen. You need to discover a convenient physical size by trial and error. With a physical size smaller than the PowerPoint display, you can click on a corner and drag the photo to full size (or any smaller size).

An obvious complication arises with photographs less than XGA resolution (i.e. width less than 1024 pixels and height less than 768 pixels). We recommend the following procedure. First, write down the pixel dimensions of your photograph. Second, change the pixel density to 72 dpi (screen resolution). That act will change the pixel dimensions, so you will then need to reset them to the original values that you wrote down. The photo will still appear at too large a physical size in

PowerPoint, but no rule of thumb exists for the perfect reduced size. That is why we do not recommend using photographs having both pixel dimensions smaller than XGA resolution.

The proper way to bring a photograph into PowerPoint is to use the Insert Picture command rather than copying to clipboard and then pasting into PowerPoint. Procedure is far more important than you might think. Although different versions of PowerPoint may have slightly different menu setups, basically pull down the Insert menu, choose Picture, and from that choose From File. Then navigate the dialog box to find the photograph to be inserted. Inserting in this way compresses the photograph, much like the perhaps familiar jpg compression. If you select the whole photograph, copy it to the clipboard, and then paste it into Powerpoint, the picture will not compress.

TIP

Insert photographs into PowerPoint instead of pasting them in.

To find out approximately how much difference procedure makes, we created three identical PowerPoint presentations with exactly one slide each. In one, we inserted a small photo (500×357 pixels) and in another we copied and pasted the same photo. The third one was left blank. The "inserted" presentation took 92 kB of storage space whereas the "pasted" one required 440 kB. The blank presentation required 16 kB ("overhead" in jargon), so subtracting that gives a comparison of 76 vs. 424 kB for the photos themselves. As photographs are read into memory when running PowerPoint, pasted photos will require about five times as much memory as inserted pictures.

Using PowerPoint

When presenting tables or figures, call attention to the data that you want your audience to focus on. If you have a pointer, use it, being careful to hold it steady so that your audience does not get distracted.

If you do not have a pointer, use words to refer to whichever column in a table or relationship expressed in a figure you are describing. Unlike written reports, in an oral presentation you cannot expect your data to speak entirely for themselves because your audience will rarely have the time to study and evaluate everything that is shown. It is your responsibility as the speaker to coach them along.

Varying the type of slide and length of projection time per slide will help maintain your audience's attention. Interspersing a series of text or data slides with photographs will help to break the monotony. You may also want to have a few summary slides in key positions towards the end of your talk in case you inadvertently fall behind and need to wrap things up sooner than you planned. How many slides you show will depend entirely on how much time you will spend on each one. A series of three to four slides showing your study subjects eating different foods or in feeding postures, for example, may require no more than 15 seconds apiece to make the point that diet or postural behavior is variable. Some speakers insert multiple photographs on to the same slide, creating a collage effect. While this is effective at displaying sequences or variation, we recommend that you consider the number of photographs that your viewers can take in at any one time. Too many photographs may be overwhelming and therefore defeat the purpose for which they were intended. Also keep in mind that animating a slide with multiple photographs, so that each appears in sequence, will take time if the image files are large. Data or text slides, by contrast, may require two to three minutes (or more) to explain. The material you are presenting and your own presentation style also factor into decisions about the number of slides you will show. Many experienced speakers use this rule of thumb:

RULE

Show no more slides than the number of minutes allotted to your talk.

Whether you bring your presentation on your own computer or on a storage device, such as a CD or USB stick, it is critical that you verify that your presentation will be readable on the equipment that your hosts can provide. Your own computer may require special attachments to connect to the LCD projector available, and you should bring these items instead of assuming they will be provided. You should also always bring your own computer's power plug and set it up before you begin so that there is no risk of running out of battery power during your talk. Almost always, the LCD projector needs to be turned on first. If you startup your own computer before it is connected to the powered up LCD, the projector probably will not be able to locate it. Be sure you know how to reset your computer display so that the image on your computer screen will be projected. You can set this in advance, but make sure you know how to make the adjustment on the spot because these settings sometimes revert to default mode at startup unless the computer is connected to a projector. We have both watched speakers nervously fiddling with controls long after their talks were supposed to begin.

If you plan to bring your presentation on a storage device, you should verify in advance that your hosts can provide a compatible computer as well as an LCD projector. It is usually most efficient, and sometimes necessary, to copy your presentation directly on to the computer that will be running it. Some computers will not even open a PowerPoint presentation from a CD directly but have no trouble once the file has been transferred to its desktop.

However you plan to transport your presentation, be prepared for the possibility that your carefully selected colors and fonts may look different when processed on another kind of computer. PC and Macintosh computers use different kinds of colors, and some fonts don't transfer correctly from one computer to the next. One of us had the disappointing experience of witnessing deep greens and a tasteful orange being projected in unappealing shades of yellow and red due to the distortion from the LCD projector. Consult your computer and software guides about the most reliable and compatible colors and fonts in advance.

Computer "slides"

The rapidity with which computerized presentations have replaced the standard 35-mm slideshow is impressive. When the first edition of this book went to press, we referred to computerized presentations as "A newer technology . . . [that might] . . . eventually replace physical slides completely, including photographs of habitats, animals, and the like." Nowadays, we can use affordable digital cameras whose images can be directly downloaded into computers and then copied into PowerPoint presentations.

Be sure that you know where your images are stored and whether you have inserted them properly into PowerPoint. Your talk should work fine either way if you are presenting from your own computer, but the link between your talk and your image files will be lost if you copy your presentation on to a storage device for transporting your talk. We recommend that you review your presentation on a different computer so that you can confirm that all images have been saved properly on your storage device before you leave your computer behind.

The equipment used to display computer slides to audiences varies, so you need to determine in advance that everything is compatible. If the presentation hall has its own projection computer, you should be prepared to bring your presentation on a CD or other storage device that is compatible with the computer you will be provided. Most of the recent versions of PowerPoint can be read interchangeably on different kinds of computers (e.g. Macintosh and PC computers running Windows®). Nevertheless, it is still worth verifying what kind of computer and software will be available and, if possible, testing out your talk on a similar type of computer in advance. You'll want to arrive early enough to load your talk on to the projection computer. Large files can take a few minutes to download, leaving your audience impatient and cutting into your speaking time. If the hall does not have its own projection computer, you may be able to plug your own notebook computer into the hall's projection system.

In most systems, the projection computer sends the information to an LCD projector, which throws the image on a screen, virtually indistinguishable from conventional slide images. In some lecture halls, the image may be sent to monitors positioned so that the audience can see the nearest one conveniently. You may be able to see your presentation from the computer screen or a speaker's monitor, so you can tell at a glance which slide is on the screen without having to turn around to view it.

Advancing computer slides is done in several different ways. First, you can set your PowerPoint presentation to advance slides automatically, after either a uniform duration or a duration that varies from slide to slide. An advantage of automatic advancing is that you can determine the length of your talk precisely and thus stay within the time limit. This option might be more attractive for a short (e.g. 12 minutes) talk at a professional meeting than an hour-long departmental seminar, but in most cases you will probably opt for advancing slides yourself. If you are controlling the show directly from a laptop, you need only click to advance slides. At lecture halls with monitors on the podium, there will likely also be a button for advancing slides. Growing in popularity are infrared "remotes" similar to those first developed for advancing old-fashioned carousel slide projectors. If all else fails (as it did once for one of us), you can prevail upon someone in the audience to sit at the laptop and advance slides as you ask for them.

If you bring your own computer or have access to a borrowed one, you can make changes to your presentation between when you leave home and arrive at your lecture. If you need to save these changes to a CD, be sure you bring an extra CD or a rewritable (RW) CD, and make sure that the computer has a writable disk drive. One of our colleagues dragged his department's computer to an international meeting with the intention of making some final revisions to his talk en route, only to discover halfway around the world that his computer lacked a writable drive and therefore none of his changes could be saved.

Other additions to PowerPoint

Besides photographs, it is possible to add other, more specialized things to a PowerPoint presentation. It is very easy to add digitized sound bites, say to give an example of an animal's call. You can set the sound to play automatically when going to a particular slide, or you can use a little button icon (in the shape of a loudspeaker) that you click to play the sound. You can also add silent digital video clips, with the same options for playing them. Finally, you can combine audio and video by playing a video clip with sound. If the soundtrack is not relevant to the video, then playing the sound will be distracting and make it difficult to talk over. Whatever additions you use, the audio and video files are called by PowerPoint rather than being loaded into memory. That means the files should be in the same computer folder with the PowerPoint presentation file itself and transferred to whatever media you are using to transport your presentation.

Other audiovisual aids

Finally, we consider four genres of other audiovisual aids. The first is overhead transparencies, which we do not recommend. Professional meetings rarely have projectors available, although usually one can be procured for a departmental seminar if requested in advance. It is often difficult to position an overhead projector so that the image is the correct size on the screen, especially in a lecture room with fixed seats. (This problem does not occur with LCD projectors because of zoom lenses, but overhead projectors cannot be fitted with zoom lenses.) Furthermore, the projector itself is between the audience and the screen, usually blocking the view of at least some would-be viewers. The image quality of overheads is not particularly good, and the speaker must fumble through a mound of transparencies and backing papers in order to do the projecting. All that said, it may be added that modern color printers allow preparation of full-color transparencies that are attractive for small, informal groups. Moreover, overheads provide a back-up in case of unanticipated computer failures or incompatibilities. Some professional societies that now assume their speakers will give computerized presentations still recommend bringing a set of overheads as a back-up just in case.

Movies are the second genre of miscellaneous audiovisual aids, but they are rarely satisfactory for meeting talks and only sometimes useful in hour-long seminars. One cannot easily show a lot of short film clips, although one continuous movie of ten minutes or less can be a useful part of a seminar, especially to demonstrate animal behavior or other biological processes in real time. Movies of course require appropriate projection equipment. The image quality in 8-mm movies is generally poor by modern standards, and even 16-mm movies can be fuzzy. Soundtracks on movies are generally very degraded by today's sound standards, and movie equipment is more prone to malfunction than equipment for almost all other kinds of commonly used visual aid.

The third genre is the modern replacement for movies, namely videotape. Videotapes have excellent soundtracks and the video can be played into a video projector and displayed on the screen, or played through monitors positioned around a hall so equipped. Despite these advantages, potential users need to be aware of compatibility problems. Two video formats, VHS (Video Home System) and 8-mm, are commonly used in the USA and can be played back with the right equipment into any modern US television set. Furthermore, two high-bias variants of these formats (super VHS, and hi-8) are now common and require special playback equipment. All four US standards, also used in Japan, are called National Television Standards Committee (NTSC) formats.

Unfortunately, the four NTSC formats by no means exhaust types of video systems. European television has twice as many scan lines, scans through them differently, and scans at a slightly slower rate than US/Japanese television. This European system is termed

Phase Alternation Line (PAL). NTSC and PAL tapes cannot be played back on each other's equipment. As if this were not confusing enough, Sony began marketing digital video recorders in 1996; because of their potential for greater compatibility with computers, digital taping has increased rapidly in popularity. Nevertheless, digital video requires yet another special set of playback equipment. In sum, the advantage of video over movies is the good sound, clear image, reliability, and ability to show many short clips with the former. Taking your video-tapes to show at a professional meeting (especially an international meeting), however, could prove disastrous because of the considerable problems in playback compatibility.

Finally, we mention the technology of the future that is already here: computer video. Special audiovisual computers or add-on equipment are necessary to download video clips from non-digital camcorders into computers, and a lot of storage space is required for video as well as a lot of read-only memory (ROM) to play it effectively. But if you have the equipment, you can easily master the software to voice over the soundtrack, add other sounds or background music, edit both the video and the audio, and create a truly slick presentation. As mentioned above, video clips can be inserted into PowerPoint presentations so that they can be shown as easily as advancing to the next slide and projected from the same equipment as the rest of your presentation. As with image files, however, you will want to be sure that you copy your video clip along with your presentation on to any storage device and that the computer running your talk has the necessary software to display the clip.

TALKS AT SCIENTIFIC MEETINGS

Giving a good 10- to 15-minute talk is a special skill. Contrary to what you might think, short talks at professional meetings are generally more difficult than long seminar presentations for many (perhaps most) speakers. You can make only one major and perhaps a couple of minor points, and your timing must be superb. Furthermore, the audience will usually be more closely familiar with your special topic than in a general seminar, putting an onus on you to be precise and clear, especially about methods and analyses of data.

Abstracts for talks

Scientific societies send out a call for papers as far as six to eight months in advance of the meetings, and even longer in advance for many international congresses. Whether you have been invited to participate in an organized symposium or will be contributing a paper on your own, you will usually need to submit by an early deadline an abstract of the talk you will present. Some societies have strict guidelines for the length and style of abstracts and require you to submit your abstract on special forms. These forms are either enclosed with the call for papers or available online, and many societies require your abstract to be submitted electronically. You should read all of the information carefully and follow the specified instructions closely. Registration forms and fees ordinarily accompany your abstract submission. Fees may be partially refunded if, for some reason, you are unable to attend the meeting.

Talk abstracts resemble the summaries and abstracts that accompany proposals and written research reports. Abstracts are usually published in the meeting's program, and meeting attendees often base their decision to attend particular talks on their reading of the abstract. For both of these reasons – as well as the fact that many scientific societies subject contributed abstracts to anonymous peer review that determines whether a talk will be granted a slot in the schedule – your abstract should be as informative as possible.

Abstracts begin with the title of your talk, your name, and your institutional address. The first sentence of your abstract should be a topic sentence that identifies the area of your research and the "why" of the study. The second sentence often indicates your research question or perspective, often beginning with "This study . . ." followed by a clear statement of what your paper will be about. The next few sentences of an abstract should indicate the source of your data, sample sizes and periods, and critical methods employed. The final few sentences summarize your most important results and should conclude with a summary statement describing your conclusions or your approach to analyzing the results. Many abstracts also indicate the funding sources for the research they will present.

It may be difficult to summarize your study's final results and conclusions if you intend to do further work during the six- to eight-month interval that usually occurs between the deadline for submitting a meeting abstract and the actual meeting date. You may want to clarify in your abstract that your results are based on analyses of a subset of your data, and that these suggest or appear to support your conclusions. In your actual presentation, you can correct any differences between what you said in your abstract and what you subsequently discovered or concluded. (Of course, some scientists present only results they have already published, but we think that practice tends to defeat the point of a meeting.)

Abstracts rarely include citations to other published works. Instead, more general statements, such as "Studies of . . . typically focus on . . ." or "Few studies have examined . . ." are made. Omitting references in your abstract does not preclude you from referring to other researchers by name in your talk. Indeed, full credit should be given when appropriate in all other forms of scientific communication.

Keep in mind that some journals (e.g. *Animal Behaviour*) have explicit rules about not accepting articles for publication if the same data have been described in an abstract that has been published in a volume with an ISBN (international standard book number) or ISSN (international standard serial number) – unique numbers assigned to publications. Ordinarily, booklets of abstracts distributed to attendees at professional meetings are personal communications rather than publications, and they do not carry registered international standard numbers. If there is doubt about the publication status of your abstract, you can limit the results you provide to general qualitative statements (e.g. "On average, males spent significantly more of their time traveling than females . . ."} instead of providing the actual descriptive statistics and analytical values. Be sure to consult the instructions for authors section of any journal to which you may want to submit your work for publication. Details on overlap restrictions for submissions to *Animal Behaviour* are provided under the subheading on the cover letter, but they may be stated elsewhere in other journals.

When you prepare the abstract, you may also be asked to indicate key words or possible session titles to which your presentation could be assigned. Careful selection of these indexes will help meeting organizers to group your talk with others on related topics into an appropriate session. Meeting participants sometimes decide which sessions rather than which papers to attend and may neglect to notice your paper if it is buried in an otherwise unrelated session.

Obtaining feedback on your abstract for a scientific meeting can be just as valuable as it is on your grant proposals and manuscripts. Allowing sufficient time to send your abstract to various people will enable you to clarify any confusing statements.

TIP

Use your word processor to make your abstract fit the box on a form for hard-copy submission. Set the left and right margins so that the text will be contained within the box provided, and space down until the first line is below the top of the box. Then type a draft of the abstract and print it on plain paper. Align the printed sheet with the form and hold them up to a window or strong back lighting; readjust the margins and the spacing down for a perfect fit. If the drafted abstract is too long, revise to shorten it, and then print again on plain paper. Continue this cycle of adjustment and revision until your printed draft fits exactly, and then print the final version on the form.

Abstracts submitted online should also be typed in a word-processing program first, where they can be reviewed carefully for content, format, and spelling, and then saved. Some online submission procedures permit you to copy and paste your saved abstract into a form; others require that you directly download your saved file. When submitting an abstract online, be sure to save a hard copy of your abstract and that you receive a confirmation that your submission was received.

Content of a talk

The brief time period allotted for most meeting talks places severe constraints on what you will have time to say and places a high premium on organization and clarity. You should plan to speak for no more than 12 minutes if you have 15 minutes, and for no more than 17 minutes if you have 20 minutes, in order to allow time for one or two questions. Session chairs are encouraged to enforce time limits and are often given alarm clocks to signal that your time is over. Failure to conclude your talk within the allotted time suggests poor preparation and may appear to be arrogant or disrespectful to other speakers and your audience.

Here is an anecdote. One of us attended a case being argued before the US Supreme Court in which the young defense lawyer was being coached by his law school mentor. Afterwards, the lawyer asked his mentor how he had done. "Fine," came the reply, "but you didn't stop when the light came on – you finished your sentence." If the chair of your session declares that your time is up, stop.

Time constraints require that meeting presentations be much more specific than a seminar presentation. You will not have time to provide as detailed an introduction, as careful a review of your methods, or as many results or conclusions as you would in a seminar. Instead, limit your talk to no more than about three points (one being the major message), and include only the most essential background and methodological descriptions to support your data and conclusions. You may have one central question that you address with three separate sets of results. It is legitimate to indicate that your talk focuses on a small aspect of a larger, more comprehensive study, but you will lose your continuity if you digress into other aspects of the larger work. Conclude with a take-home message: a statement of the central point made by your study. The final slide of a talk usually includes acknowledgments to advisors, collaborators, and funding sources, although some speakers make the acknowledgments at the start of their talks. You may want to call attention to key individuals or specific organizations listed. Close by thanking your audience for their attention.

Slides

The greater time constraints and narrower focus of meeting talks, compared with seminar presentations, also mean that you will need to be more selective in your use of slides. The earlier comments on slides apply here as well, but some bear repeating in this special context of a meeting talk. The maximum of one-slide-per-minute should be violated only if there are photographs of animals or habitats that will be on the screen only briefly. A typewritten double-spaced page of text usually corresponds to about 1.5 to 2 minutes of read or spoken material. Writing out your talk, even if you will ultimately talk it through instead of reading it, may help you to identify how many slides you will need. With experience, you may discover, as we have, that it is easier to develop your talk from a list of the slides you intend to show. For example, what are the essential points you need to make about your study subject while you are projecting a slide of it during your talk? There will not be time to provide your species' entire natural history; rather, you will need to limit your comments to those aspects of your species that pertain directly to the subject of your presentation.

Preparation of slides for meeting talks is similar to that for seminar presentations. Compressing too much information into a limited number of slides will be as unintelligible to your audience as overwhelming them with so many that you must race through the projection of each. Remember the rule of thumb: a maximum of eight lines on a slide of text, or five rows and three columns of tabulated material.

Familiarizing yourself with the room you will be speaking in should be standard practice. Arrive at your session before it begins and check out the podium, making sure that you also know how to work the lights, electronic pointer if one is provided, and microphone.

Most scientific meetings now provide details on the types of computer and software they will provide, specify how to store your presentations, and how far in advance of your talk it will need to be delivered. Because you will not be running your presentation from your own computer, you will want to save and test your presentation in advance on the same kind of computer that will be in use at the meetings. Once you arrive at your meeting site, make sure you know where to bring your saved presentation for downloading on to the operating computer. Some large meetings request a copy of all speakers' presentations as much as a day in advance of their scheduled speaking times, while others ask that speakers arrive an hour or so in advance of their sessions in order to download their presentations on to the computer stationed in that room.

Practice

Meeting talks are just as important to practice in advance as seminars, especially because of the strict time limits imposed. You may need to practice a meeting talk more than once as you shorten or modify uneven parts. If other colleagues from your institution will be presenting papers at the same meeting, then you may agree to set aside a block of time to rehearse one another. Listening to and observing how others deliver meeting talks, even in a practice session, can help you learn how to present your own. Ask your colleagues to be critical and to watch for any confusion in your content or annoying habits you may be unaware of. Have someone time your talk, ideally noting where you are in your talk at five-minute intervals so that you will know what to eliminate or reduce if you exceed your limit.

Style

Meeting talks tend to be more formal than seminar presentations, in part but not only because of the limited time available. Many speakers opt to read their talks at scientific meetings in order to ensure that they state precisely what they had intended to say and to prevent themselves from unintended digressions in the more limited time available. Others prefer a more conversational style even when giving a paper at a meeting. In either case, prior practice is essential. Even reading a talk requires that you know where to pause, inflect, or emphasize in the text and how to keep track of your place in the text while advancing the slides and any animations they may include. A spoken rather than read talk also requires you to know when to advance your slides, and, as in the case of seminar presentations, good slides placed appropriately make better meeting talks.

Written communications can rely on the reader to work through a confusing passage or to go back in the text to recover a lost train of reasoning, but an audience listening to a paper being read does not have these options. An audience will not see the paragraph breaks or section headings, so short sentences and clear transitions between passages become essential to accompany the argument.

Listening to a speaker "talk" the paper instead of reading it can be easier because the appropriate pauses, inflections, and emphases are more likely to come naturally. The drawbacks of a spoken rather than read meeting paper are that the material may not be stated as precisely or concisely as it would be in writing. Practicing your talk will permit you to identify which parts of it you may have difficulty articulating, and repeated practice will help you to smooth out these parts.

POSTERS AT SCIENTIFIC MEETINGS

Many scientific societies actively encourage poster contributions, both to alleviate the problems of scheduling multiple overlapping paper sessions and to stimulate greater communication between presenters and attendees. Your decision as to whether to present a poster or oral paper may depend on the material you intend to communicate. A few meetings now accept contributed papers only as posters, however. Then, from the abstracts submitted, they may invite a small group to make oral presentations instead, if they so choose. Either way, it has become increasingly important for a scientist to be able to make an effective poster presentation.

Abstracts for posters

Most scientific meetings require poster abstracts to be submitted at the same time as abstracts for oral papers. Some societies ask you to indicate whether your abstract is intended for one or either type of presentation. Guidelines for poster abstracts are identical to those for oral papers unless indicated otherwise in the submission materials (see our earlier comments on abstracts for meeting talks).

Content of a poster

Presenting a poster is an ideal opportunity to get abundant feedback from others in your field about preliminary data and ideas, and to meet other scientists in an informal context. Each poster is usually allotted at least a full morning or afternoon for display, and you are encouraged to stand by your poster for at least part of that time in order to answer questions or elaborate on the material you are presenting. Some meetings now have specific poster sessions during which no concurrent activities occur. Usually, authors are required to be at their posters during such dedicated sessions.

Posters work best for straightforward results. They usually include a panel for the title, author, and author's affiliation, and text panels that introduce the research, outline the methods, and summarize the conclusions. Some meetings encourage or require the poster to contain in an upper corner a small photograph of the senior author. Results may be presented in tables or figures, with informative captions that summarize the data. Complex theoretical arguments may be easier to explain in an oral paper than a poster, but most empirical studies are suited to either.

The submission guidelines will indicate the approximate size of the display area each poster will have. Usually no more than 12 US standard 8.5×11-inch panels or A2 sheets will fit. Compressing too much text or too many data panels into a poster may discourage people from stopping to examine it.

An increasingly common trend is to prepare the poster as a whole and print it on one huge sheet using a special computer-driven plotter. At a recent meeting attended by one of us, approximately 90% of the posters were of this type. If your institution does not have the hardware available to make such posters, they can usually be made at special commercial copy shops in large cities and college towns. These posters have several advantages. One is that they can be carried on airplanes and in other vehicles rolled up or contained in a cylindrical map case for protection; thus, you should never lose part of the poster. Another advantage is that they require a minimum number of tacks or tape in order to mount them at the meeting. Yet another advantage is that after the meeting, you can take the poster back to your home institution and mount it on the wall of a hallway for peers to see.

A viable strategy for a good poster presentation mimics a classical newspaper story, which has an informative headline (and perhaps subheadline), followed by an encapsulated story in the first paragraph (in effect, an abstract), and then the expanded story. Your poster title is the headline and you can add a longer and more informative subtitle. Abstracts are not required on poster displays but can be used effectively in parallel with the lead paragraph of a newspaper story. Then the rest of your poster is the full story. A final, smaller-font panel with acknowledgments is also appropriate for a poster presentation.

Illustrations

The success of poster presentations relies on attracting people to read and talk about your paper. In a room lined with many posters, esthetics become as important as legibility. All text and figures should be printed large enough to read at a distance of about 3 feet (at least arm's length). The guidelines for poster presentations at the American Association of Physical Anthropology meetings recommend that the title be legible from 8 feet and the rest of the poster from 5 feet away. Boldface 18-point or larger fonts generally work best for text, with at least 30–36-point fonts for major headings. In crowded poster sessions potential readers may not be able to get even within 3 feet of your poster, so put the most important printed material high on the poster and in large type.

Photographs, ideally about 8×10 inches in size and accompanied by legible captions, can also be visually attractive and informative and help to break up the monotony of a poster display. Like oral presentations, most posters include a mixture of text, data tables or preferably figures (i.e. graphs), and one or more photographs. Posters can be constructed in an illustrator program, such as Adobe Photoshop or Adobe Illustrator®, or in PowerPoint and then saved as a pdf document. You can set the size of your document to match that of your poster, which can then be printed out on a single large sheet at facilities with appropriately sized and configured printers. Be aware, however, that settings may change in this process, and incompatibilities, especially between Macs and PCs, are not uncommon.

The title should be large and extend across the width of the poster to capture attention. Like written papers, figures should include legends and be numbered and located according to their reference in the text. Posters also usually include a brief list of references cited and should always conclude with appropriate acknowledgments. Play with your poster panels, which are the equivalent of slides, to determine the most logical sequence to display them in, and sketch out in advance a diagram of where each panel will go.

Mounting

When the first edition of this book went to press, most meetings' posters were still being mounted on colored poster board. Nowadays, computer-generated posters, printed as single large sheets, are the established norm. Once you have your poster presentation ready in digital format, you can take it to a facility on your campus or a local copy shop with a large enough printer to print it out. The cost of printing a standard sized meetings poster varied from about $15 to $100 in 2005, depending on the facility and the services available. Mistakes can be costly, so previewing your poster's layout and editing all text carefully in advance will save you money in reprinting or the embarrassment that comes from standing under a poster with an obvious typographical error. Printing your poster far enough in advance will also give you time to correct any mistakes and reprint it again if necessary.

The single poster sheets are flexible and can be rolled up for easy transport in a cardboard or plastic tube carrier. You will want to release your poster from its carrying container when you get to your destination so that it has time to flatten out before you need to mount it.

Thumbtacks or strong pins are usually used to attach the mounted panels to the display board. Most meetings provide these materials, but it may be worth the extra investment to bring some along in case of shortages. Colored pinheads that match the poster are less distracting than bright silver or gold tacks.

It is always a good idea to confirm what materials will be required for mounting your poster if this information is not provided along with the meeting instructions. One of the reviewers of this book described a meeting at which the conference center provided hard panels with a cloth covering intended for Velcro® tabs on the poster but failed to inform poster presenters in advance. As with oral presentations, be sure you are fully aware of any materials or aids you will need in order to avoid unnecessary and potentially disastrous last-minute surprises.

Exhibiting skills
Find your designated display spot and arrange your poster at the designated time. Stand back and make sure that the poster is level.

As people arrive to look at your poster, be sure that you are standing to the side instead of in front of any part of it obscuring their view. Do not stand so far away that your association with the poster is unclear. You may offer to walk visitors through your poster, elaborating on each panel and explaining your study and your results as they follow along. The more interactive you are with your audience, the more feedback you are likely to get. Crowds attract crowds, and an active exchange between you and a visitor will arouse the interest of other people nearby, who may join in to learn about your research and to ask questions that may stimulate your next study. Nevertheless, be aware that many visitors are shy or do not wish to engage in conversation because they have so many posters they want to see that they must read quickly and then move on. Therefore, do not accost passersby but do have your meeting name tag displayed prominently and be prepared to respond promptly if someone appears to want to talk with you.

5 How to write a curriculum vitae

Personal information
    Standard personal data
    Optional personal data

Education
    Degrees
    Coursework

Honors, awards, and similar recognitions
    Awards and honors *sensu stricto*
    Prestigious professional service

Grants

Publications

Professional talks

Teaching

Miscellaneous optional information
    Memberships and service
    Certifications, qualifications, and skills
    Employment and other experience
    Research summary

The writer of his own life has, at least, the first qualification of an historian, the knowledge of the truth.

Samuel Johnson (1709–84)

A curriculum vitae, affectionately known as a c.v., is a summary of one's academic career and qualifications, usually prepared by an applicant seeking employment or other support. The commercial world tends to use the French term résumé for the equivalent document. *Curriculum vitae* means literally the course of life (in Latin), and it intends to be a short summary, although modern c.v.s can sometimes be quite lengthy, depending upon the specific purpose for which they are drawn up. Although most of a c.v. consists of itemized lists, some general exposition may be involved; much of Appendix A on writing clearly therefore applies.

There is no prescribed format for a general c.v., although its contents are reasonably standard. For specific purposes, as in a tenure-review document, a university or other body may require a c. v. in special format. We emphasize the contents that typify a c.v. drawn up by a new doctorate seeking a postdoctoral position or assistant professorship, although we also include mention of sundry items that are often more applicable to someone at a later stage of his or her career. As it may not be obvious why certain items are desirable to include in a c.v., we offer our explanations as to their inclusion. However, we urge you to seek feedback on your c.v. as you prepare it for different purposes and whenever you substantially revise its format or content.

People seeking jobs in commerce often tailor their résumés for each different application; these kinds of résumé tend to be condensed into one or two pages. Academicians, by contrast, tend to have one c.v., or in some cases one "complete" c.v. and a shorter version that is suitable for transmission by electronic mail or fax. Nevertheless, there may be occasions for which academic applicants wish to have two or more versions. For example, a new doctorate applying simultaneously for postdoctoral research positions and professorial appointments may want to omit or shorten teaching experience in the c.v. accompanying the applications for the research positions while including instructional history when applying for the teaching positions. The important thing is to provide the information relevant to the position sought.

Generally, research prowess is the most important qualification of an academic. Applicants for postdoctoral work will be evaluated almost completely on their research promise, and those seeking a professorial position will be judged strongly on research along with promise as a teacher. Therefore, presenting your research accomplishments clearly and fully in your c.v. is important.

We also encourage you to document your accomplishments throughout your academic career. In addition to a portfolio of manuscripts you review (discussed in Chapter 4), it may be useful to save letters of commendation or acknowledgment for professional services, or any other hard evidence of your professional activities. Saving a copy of all evidence of your teaching activities, even when as a laboratory instructor, discussion leader, or teaching assistant, and any notes about your efforts to enhance student learning at all levels, is also important. These materials might include any course syllabi, sample tests, or laboratory assignments that you developed, as well as both formal teaching evaluations and unsolicited student letters. These documents should not accompany your c.v. unless specifically requested, but they will often be required by promotion and tenure committees. Getting into the habit of filing supporting materials early on in your career will help you pull them together when you are expected to provide them.

PERSONAL INFORMATION

Although the division is arbitrary, it is convenient and natural to discuss first standard personal information that probably should go into every c.v. and only then consider some items that you may or may not wish to include. Personal information usually comes first, regardless of the format used for constructing a c.v.

Standard personal data

It is standard practice to provide your full name, place of birth, permanent residency or citizenship if not obvious from your place of birth, and current academic address (including telephone number, fax number and email address). You may also opt to include your social security number, but be careful about including it on a c.v. that may ultimately be posted on a website or transmitted widely over email – unfortunately, identity theft has become quite common in the USA. The usefulness of some of these personal items may not be obvious, so we consider them here before discussing information that is more optional.

A principal use of much of the information listed here is for accounting purposes of an institution that will issue you a check to reimburse travel expenses or pay a small honorarium for your giving a departmental seminar or other talk. Women fearing possible job discrimination based on sex might be tempted to provide only their initials instead of full first name, but even if the fear is well grounded letters of recommendation and other information will reveal the applicant's sex. Furthermore, accounting and tax procedures usually require full names before a check can be issued. Similarly, the federal government in the USA requires that disbursement of potentially taxable income be accounted for by home (not business) address plus social security number. Failure to include the needed information in your c.v. could delay a reimbursement check while the institution contacts you to obtain the information, but this delay may be acceptable if you are concerned about the extent to which your c.v. will be distributed.

If you were born in the USA and provide the place of your birth in your c.v., there is usually no need to state explicitly that you are a US citizen. If you are not a US citizen, or you were born outside of the USA, then it is a good idea to provide your nationality or the country in which you hold permanent residency. Some governmental programs and private foundations cannot expend certain funds in payment to nationals of other countries. Conversely, some types of fellowship are restricted to people who are not US citizens. Furthermore, universities and other institutions may have to file certain papers with the federal government when hiring people who are not US citizens. It will usually save you later hassle if you provide citizenship information along with other personal data in your c.v.

Optional personal data

Beyond the standard data, one could add personal information to a c.v. according to individual preference or circumstance. For example, you may wish to include years of military service (if any). Without accounting for those years, it might appear that you are attempting to conceal something about your career.

As another example, including your middle initial or name can help prevent confusion with other people with similar names. You may even wish to adopt a distinctive form of your name.

Birth date is another optional inclusion. Some people may be reticent to include it for fear of age discrimination in hiring, although everyone in the USA is protected by law against this. To discriminate against anyone based on their sex, age, or one of a variety of other attributes deemed irrelevant to employment is illegal.

Marital status is irrelevant to academic qualifications, and employers in the USA are forbidden to use such information in hiring decisions. Therefore, it is definitely optional to include such data in a c.v. There may be professional reasons to include information on partners or children, but many of our colleagues consider their families to be personal.

EDUCATION

Regardless of the stage of your career, at least a minimum summary of your educational background is a mandatory inclusion in the c.v. Earned (as opposed to any honorary) academic degrees are the most important element, but other information may also be included, depending upon the stage of your career.

Degrees

Most people reading this book will have (ultimately) at least two degrees to document: a bachelor's degree and a doctorate. Many will also have a master's degree, and a few may have two degrees at the same level, some possible patterns being master's degrees in both biology and some other area such as statistics or chemistry, or a Ph. D. plus an M. D. or D. V. M. The c.v. should list as minimum information the degrees by initials, the year each was awarded, and the awarding institution.

Ordinarily, abbreviations of degrees do not require explanation. Harvard's A. B. will be recognized as a bachelor's degree like the more common B. A. or B. S., and Oxford's D. Phil. is obviously the equivalent of the more common Ph. D. If an abbreviation strays too far into the unfamiliar, however, a phrase of explanation may be useful. In the same vein, the location of an awarding institution usually need not be provided unless confusion could arise. For example, Miami University and the University of Miami, despite the difference in official names, are liable to be confounded, so it is wise to specify Ohio or Florida. Similarly, Washington University (in St Louis) and the University of Washington (in Seattle) could easily be confused.

Two further elements of specification are useful although not generally considered mandatory: the major (and, if applicable, minor) fields of specialization and the name of the advisor or supervisor. One of us specifies biology as the undergraduate major and zoology as the major field of the doctorate, with a psychology minor, thus summarizing zoology as the educational core with some background support in botany and psychology.

The name of the advisor, especially the thesis advisor of the highest earned degree, should probably be considered a nearly mandatory inclusion on the c.v. One of us once sat on a search committee that noticed the thesis supervisor's name was not specified in the c.v. of a candidate. Nor did the letters from any of the candidate's referees indicate that they had supervised his research. This situation prompted a telephone call to the awarding institution and eventually contact with the major professor, whom the candidate had not asked to write a letter of support apparently because the two individuals did not get along well. The committee's having discovered the facts in this roundabout manner was an unnecessary irritation that could have been avoided by being more straightforward on the c.v. It is better to make the best of a bad situation through full divulgence than to exacerbate the situation by attempting to conceal it.

Coursework

It is usually not necessary to list the formal courses of your higher education because employers normally require an official transcript from the institutions at which the work was done. Nevertheless, you can optionally provide an organized summary if it is important in an employment niche to which you aspire or some other reason obtains. For example, a person whose degrees are in biology may wish to point out a strong background in some supporting area such as statistics and computing, especially if the coursework is scattered through long transcripts and not obvious as a coordinated program of study in addition to the primary field.

HONORS, AWARDS, AND SIMILAR RECOGNITIONS

Do not be bashful about providing an objective record of recognitions you have received. Many people place this section of their c.v. third in rank after personal data and educational summary. Regardless of where you place it in the c.v., evidence that others have recognized quality achievement in your academic work and promise is part of your qualifications.

What constitutes a recognition is open to interpretation because the academic world is so wonderfully diverse. Many people separate out into a distinct section awards of funding such as scholarships, fellowships, grants, and contracts. If these are few, however, they could just as easily be included with other honors and awards. Similarly, appointments and elections to offices of professional societies may also be listed separately if several or included with other honors and awards if the number does not merit separate listing.

reference more difficult to locate in a library and seeming to imply that you are always senior author, even when that is not the case.

The c.v. references may include those in press, as is the case with journal manuscripts (see Chapter 3), but can go beyond this to include manuscripts submitted and currently under editorial consideration. New doctorates in particular may wish to make such inclusions, indicating that they have been active in preparing their dissertation results for public scrutiny. Nevertheless, it is mandatory to make clear that such manuscripts have not been accepted for publication but merely are "submitted" or "under consideration." Furthermore, listing of such manuscripts entails the risk that they will not be accepted, thus possibly providing a minor embarrassment in the future.

PROFESSIONAL TALKS

Most graduate students and postdoctoral associates will probably want to list departmental colloquia and contributed papers presented at professional meetings. The departmental seminars will initially arise mainly from job talks at institutions where the candidate is on the short list of those under serious consideration. The listing of research presentations serves a dual purpose. First, it provides evidence for public-speaking experiences that are relevant to preparations for offering one's own lecture courses. At the same time, the listing shows a history of promulgating research results, especially considering the usual delays before a finished manuscript sees print in a journal. Give the title of your talk, the date of presentation (month and year), and the occasion and place.

More senior researchers will probably wish to be selective in their listings, so as not to appear to be padding out the c.v. with minor items that are more relevant for new doctorates. At some later stage in one's career, departmental colloquia may still be relevant, but the emphasis should be on invited talks, such as symposia contributions, keynote addresses, and plenary talks at scientific meetings.

TEACHING

Applicants for academic positions should include in the c.v. a summary of their instructional history. In some cases, this history can begin with upperclassmen years of college where programs exist in some institutions for senior teaching apprenticeships, which provide experiences much like those of graduate teaching assistantships. More commonly, one's instructional career begins with graduate teaching assistantships but in some cases can go beyond this, where graduate students overseen by a faculty member can offer their own small courses or seminars for undergraduates. Postdoctoral research associates may also have the opportunity to offer a seminar carrying academic credit for graduate students. Junior faculty will of course begin listing the courses they have taught, phasing out the listing of those for which they were graduate teaching assistants earlier in their careers.

Whatever the nature of instructional involvement, it is common to list the course by its title (e.g. History of psychology) rather than by its institutional number (which will mean little to readers at other institutions). The year in which the teaching experience took place is also useful documentation, and some people include the number of credits and the size of the class as evidence of the extent of the experience.

Some people include in their c.v. a free-form statement concerning their teaching goals. We authors are of two minds about the advisability of such an inclusion. Teaching philosophies differ greatly among academics, so one risks provoking a negative reaction in readers whose approach to instruction does not match that of the job candidate submitting the c.v. New Ph.D.s in particular rarely have sufficient experience to grasp all the subtle and often conflicting aspects of teaching. Therefore, it may be better not to make a premature statement in the c.v. Even listing the courses one feels qualified to offer might open the door to problems, for teaching needs differ between departments and over time in one department. It is true that search committees at research universities are increasingly interested in the teaching philosophy and experience of candidates for their jobs – so we do recommend that you are prepared to provide a specific statement concerning instructional plans and philosophy if and when asked.

MISCELLANEOUS OPTIONAL INFORMATION

Many people may wish to include other kinds of information that are relevant to professional qualifications and promise. The heart of a c.v. is personal information, educational history, and honors, and then the specifics of teaching and research experience and accomplishments. It is impossible to anticipate all the kinds of miscellaneous data that might also be relevant for inclusion, but in this section we offer some examples of things that could go into a c.v. How this material is organized is largely a matter of personal preference and convenience.

Memberships and service

Graduate students in particular may wish to include a list of the societies to which they belong, as well as general service commitments to the academic community. Elected memberships are generally considered honors (see the previous section), but merely paying dues to a professional society shows an active interest in one's chosen field. Often in behavioral ecology and related fields, researchers may belong to two types of society: those according to discipline, such as the Ecological Society of America or the American Association of Physical Anthropology, and those devoted to all aspects of biology of a particular animal group, such as the American Society of Mammalogists or the American Society of Primatologists.

Most developing scientists will become involved in tithing some of their time for service to their professional societies or their home academic institutions. For example, many professional societies ask for volunteers to serve on committees and assign willing people according to their experience. Thus, graduate students may be assigned to relatively unimportant and less demanding committees but this assignment is a stepping stone to wider responsibilities.

Graduate students may also be active on campus, in some departments serving as student members of departmental committees or as officials in a graduate student organization. Such service may be listed in the c.v. as evidence of commitment.

Certifications, qualifications, and skills

Many people will have qualifications that will not show up on a transcript because they were not achieved through formal coursework. Some examples of formal qualifications might include teacher certification for secondary schools and certifications especially relevant to field workers and marine biologists in particular, such as Red Cross certifications in lifesaving, first aid, or cardiopulmonary resuscitation, and scuba qualification from the Professional Association of Diving Instructors (PADI) or an equivalent program.

One may also possess relevant skills that are not formally certified by training courses or acquired through coursework in higher education, such as competence in computer languages. Other people might have field-relevant skills, such as the ability to make major repairs on vehicles, mountain-climbing experience, and so on. Language skills may also be important, especially for those who aspire to do overseas field work. Many people acquire language competency at home by virtue of having relatives whose native language is not English. Even if language competency was obtained through formal education, it might be useful to make mention of it in the c.v., as courses could be missed by those reading the transcripts.

Employment and other experience

For government and industry jobs, one's employment history is a principal qualification, usually featured near the head of a résumé. However, for academics (especially graduate students), there will often be no history of full-time employment. Nevertheless, some people will have had full-time jobs between college and graduate school that are part of their relevant qualifications, such as teaching in secondary schools or working in industry as a computer programmer. Furthermore, some types of summer employment or part-time jobs during college or graduate school may also be useful to list, such as working as a nature counselor at a summer camp, a rafting boatman, lifeguard, or ski patrol.

Another kind of relevant inclusion is field experience. College students often find volunteer or low-paying summer jobs as field assistants on research projects. If these experiences are overseas, then they constitute even stronger background.

Research summary

Many people write a succinct statement of their research interests and accomplishments for inclusion in their c.v. The summaries can be useful for search committees wanting an overview of the candidate's research and for hosts preparing their introductions to talks by the person providing the c.v. before a campus visit. There is a hazard in preparing a free-form statement of research interests, accomplishments, and goals; namely, if they are not carefully written, such statements can appear vaunting. Unlike most of the c.v., which consists of listings of fairly objective data, these more subjective statements attempt to summarize in plain English the main features of a research program, including where it has come from and where it is anticipated to go. Like the teaching statement mentioned previously, including a narrative describing your research interests is a personal decision unless it is specifically required.

Appendix A How to write clearly

Generalities

The three primary rules

Structure
  Subsections, blocks, and paragraphs
  Sentences
  Word order

Numbers
  Units
  Precision

Abbreviations and animal names
  Latin abbreviations in exposition
  Abbreviating scientific names
  Initial capitals in avian names

Using signs, symbols, and marks
  The hyphen (-) and the dash (–)
  The number symbol (#)
  The exclamation mark (!)

Difficult inflections
  Matching number of subject and verb
  "Noun string construction grammar"
  Adjectival degrees
  Those awful subjunctives

Problematic pairs

Using a word processor
  Headers, footers, and line numbers
  Spell checkers

Care should be taken, not that the reader may understand, but that he
must understand.

Quintilian (*c.* 35–95)

This appendix could be called *"Some tips on* how to write clearly"
because it cannot – and does not attempt to – replace detailed guides
on the subject of English exposition. We focus instead on issues
and problems that arise frequently in scientific writing, as in grant
proposals (Chapter 2) and research reports (Chapter 3).

Many writing guides are biased towards being prescriptive
rather than descriptive. That is, they prescribe what writers should
do rather than describing what writers do. Part of this bias results
from the very nature of such guides; it is easier and shorter to say what
to do than it is to explain why or to show that good writers do things
in the way prescribed. The other part of the bias results from selection: those who write guides tend to be philosophically prescriptive
about language.

It is with some trepidation that we offer these notes about how
to write clearly because they inevitably appear, and often are, prescriptive. English has become the world language because it is forgiving in so many ways. Any writer can violate many of the so-called rules
and still be understood. Truly good writers see through awkward
prescriptions. For example, everyone knows the following "rule:"
never use a preposition to end a sentence with. Winston Churchill,
perhaps the greatest master of the English language since William
Shakespeare, is said to have countered with "This is an impertinence
up with which I will not put."

Furthermore, language evolves, and today's "rules" may seem
silly and old-fashioned by tomorrow's readers. There are no grammar
police, and the only penalty for breaking a rule is possibly to confuse
your reader. So consider our prescriptions as mere suggestions, consult
other authorities, and above all follow the examples of good writers.

meaning clearly. Supposedly, only Chinese among the major languages of the world leans so heavily on word order to convey intended meaning. As users of English, we learn not to split infinitives but we rarely stop to consider other aspects of word order because the words flow so readily from our mouths and pens. Nevertheless, a couple of cases are worthy of consideration.

You may have forgotten about the prohibition of dangling participles, so here is a reminder. A participle, usually in a subordinate clause, that lacks any clear relation with the subject of the sentence is said to "dangle." For example, "Approaching Isle Royal, the wolves came into our view" lacks clear specification concerning who was approaching. It would be fine to write "Approaching Isle Royal, we got our first view of the wolves." Fixing the dangle requires a little more than reordering the words, but the problem is fundamentally one of logical relations among key words.

Even if you never dangle a participle, you have probably at some time written something like this: "I could only see the bird when it sang." Literally that means "When the bird sang, I could see it but not hear it or sense it in any other way." Change the order, and the words convey their intended meaning: "I could see the bird only when it sang."

NUMBERS

Writers sometimes have trouble expressing numbers, but there are simple guidelines. The general rule of exposition is to spell out integers up to ten and use numerals for all other numbers. However, scientific writing in particular entails further distinctions. Any number that is followed by units is always expressed as a numeral, thus: "We observed nine monkeys within 9 m of one another." Furthermore, numerals cannot begin a sentence, as noted previously. General writing sometimes mixes numerals with words, as in "The debt was $46 million" instead of "$46 000 000" or "forty-six million dollars." Such mixtures rarely occur in scientific writing because of special ways to deal with expressing significant digits (see *Precision* below).

Units

The writing of units for quantities is dictated by the Système International (SI) system. This system is an international standard that enjoys widespread, although not universal, adherence. The SI units are based on the metric system, and each has a sanctioned abbreviation, thus: "A chickadee weighs 9 g and a titmouse is larger." There is always a space between the numeral and the unit abbreviation that follows it, and there is never a period after the abbreviation (unless it concludes the sentence, of course). In some cases, the SI system provides alternative abbreviations (e.g. "s" or "sec" for seconds). The unit of absolute temperature is the kelvin, abbreviated K (without the degree symbol). Because the SI definitions reserve C for the coulomb, temperature in celsius must use the degree symbol (°C). Nevertheless, the American Standard Abbreviations for Scientific and Engineering Terms uses C (without the degree symbol), and this is followed by many journals in the biological sciences. Some journals modify the SI rules to prevent common confusions. For example, the abbreviation "l" for liter so resembles the numeral 1 that many journals spell out "liter." The use of the abbreviation "L" is not sanctioned by the SI or the American Standard, which use L for the unit lambert. However, some journals do use L for liter.

So-called English units are generally avoided, but they may be used if followed parenthetically by the SI equivalents; even then, the English unit may not be abbreviated. For example, it is accurate to express tape speed in reel-to-reel recorders as "7.5 inches/s (19 cm/s)" because the industry standard is actually in inches. In this case, the SI equivalent is only an approximation. The exact equivalent of 7.5 inches is 19.05 cm, but the latter value has more digits than are significant in standards of tape speed, so 7.5 inches/s is the more accurate way of reporting the speed.

Precision

Numbers in science are written so as to indicate their precision by showing only their significant digits. Hearkening back to the earlier example of $46 000 000, the number written thus technically means "the value I intend to express lies between 45 999 999.5 and 46 000.000.5." Put differently, the value is asserted as lying closer to 46 000 000 than to either 45 999 999 or 46 000 001. Writing "$46 million" indicates in general writing that the value is between $45.5 million and $46.5 million, which is to say that the value is closer to $46 million than to either $45 million or $47 million. Science expresses such quantities in exponential notation: 46×10^6, meaning that the value lies between 45.5×10^6 and 46.5×10^6.

Exponential notation cannot cure all ills in expressing the precision of a number. If the second digit itself is uncertain – say, the value of the foregoing example lies between 45 million and 47 million (instead of more narrowly between 45.5 million and 46.5 million), then expressing the number as 4.6×10^7 (instead of 46×10^6) helps to suggest that uncertainly. The two expressions 4.6×10^7 and 46×10^6 are numerically equivalent, but the first form subtly suggests a wider range within which the true value may lie. The alternative of writing 5×10^7 is worse because it implies that the intended number lies between 45 million and 55 million, which is true but unhelpfully so, because the range of uncertainty is too large. Therefore, if the precision of a number is important to the topic but cannot be expressed adequately by exponential notation, one must resort to explicit statement: for example, "the value is $45–47 \times 10^6$" (or similar expression).

Note that the precision of a number is not the same as its accuracy. An astronomer might express a value as 4.6×10^7 light-years to indicate the precision with which some instrument measured the distance. The true distance, however, could be completely out of the range of that precision. A value can be precise without being accurate. The distinction between precision and accuracy is not always as clear-cut as some textbooks imply; that subject is somewhat "philosophical" and beyond our present scope. Nevertheless, here is an analogy (supplied by our colleague Robert L. Jeanne) that may help you to remember that precision and accuracy are different concepts. A gun that shoots a tight cluster of bullets is precise, but if the cluster is not centered on the bull's-eye, then the gun is not accurate. Conversely, a gun that spreads the cluster out more is less precise, but if the cluster is centered on the bull's-eye, then it is more accurate.

Units provide an alternative to exponential notation for expressing precision. Suppose you measured some length to within a centimeter and obtained a value of 104 cm. You could express it thus, or as 1.04×10^2 cm, or by switching units as 1.04 m. In this way, units that are related exponentially can provide alternative ways of expressing the same precision. In general, it is best to avoid exponential notation when the same precision can be expressed by choosing appropriate units. If a number less than one is expressed in decimal notation, be sure to include the leading zero so that the reader notices the tiny decimal point and thus does not mistake the value (e.g. write 0.96 m, not .96 m).

Common sense must be used in expressing values that are obtained from calculations. For example, if you made a field measurement from point A to point B with a tape yielding 1.04 m and another measurement along the same straight line from B to C paced off at 27 m, then it is evident that the summed distance from A to C should be expressed as 28 m (not 28.04 m). Averages, however, have one more significant digit than their component measured values. Thus, three feeding bouts measured at 37 s, 12 s, and 46 s would provide a mean value expressed as 31.7 s. The actual numerical calculation yields $31^{2/3}$ s (31.6666 . . .), which is rounded to express the precision.

Finally, rounding a fraction that has 5 as the final digit provides an obvious problem, solved arbitrarily but conventionally. As "x.xxy5" means the number lies between "x.xxy45" and "x.xxy55," it could lie closer to "x.xxy" or "x.xx(y+1)," and we do not know which is the case. One could randomly round up or down to achieve an arbitrary split, but the convention is to round to the even digit. For example, the number 3.75 is rounded to 3.8 whereas 3.65 is rounded to 3.6. If digits are known past 5, the rounding will always be up, because the value is more than halfway to the next digit: for example, 3.75001 is rounded to 3.8 and 3.65001 is rounded to 3.7 (not 3.6). Computers do not necessarily follow these rules of mathematical expression, so the author needs to be aware of the problem.

ABBREVIATIONS AND ANIMAL NAMES

Biologists encounter Latin abbreviations of two types: those of general exposition and those relating to Latinized scientific names of plants and animals. There was a time in the USA when children routinely learned Latin in public schools, so that abbreviations in writing were used frequently and understood widely. With the loss of classical education, modern readers and writers often have difficulties with even the commonest Latin abbreviations. Furthermore, with the increasing condensation of traditional whole-organism zoology and botany in college curricula, students are rarely trained in biological nomenclature and its rules. We can merely touch upon that subject here (specifically, abbreviations used in Latinized binomials for species) and add a section about the use of initial capitals for common names.

Latin abbreviations in exposition

Chapter 3 discusses some special Latin abbreviations used in bibliographic citations (*et al.*, *ibid.*, *op. cit.*, *loc. cit.*), so only more general abbreviations are discussed here. Latin phrases and abbreviations are not as common in writing as they once were, and some scientific journals have gone so far as to forbid many of them. Even if you do not use them in your own writing, you need to be able to interpret them correctly as a reader.

Modern writers and readers sometimes confound two of the commonest Latin abbreviations, e.g. and i.e., so get them right! The first stands for *exempli gratia* (for the sake of example) and the second for *id est* (that is). Journals differ in whether they set these and other Latin abbreviations in italic. An unnecessary duplication, considered distinctly incorrect usage, is to begin a list with e.g. and then end it with etc. (*et cetera*). If the list is of examples, then use either but not both; if the list is not of examples, then only etc. will do.

Another commonly misconstrued Latin abbreviation is cf., which does not mean "see" as in "cf. Table 1" – a common misuse in scientific writing. The abbreviation is for the Latin *confer* (which means "compare"). and should always imply a contrast or disagreement – or, in the hands of a sensitive writer, even an unexpected similarity. But it is both incorrect and pretentious to use cf. as a synonym for the English word "see" (in the facile sense of refer to or consult). Also, note that the abbreviation is cf. (not c.f. as sometimes miswritten).

Still another commonly encountered Latin word is *sic*, which has nothing to do with canine behavior but means "thus" or "so." The word is almost always restricted to an insertion, in square brackets, within quoted text in order to indicate that the original was quoted correctly. The commonest use is to denote an error in the original for which the quoting author does not want to be blamed, as in "Darwin and Wallace published in 1958 [*sic*] the first papers proposing natural selection." Less commonly, an author may mark any paradoxical or surprising word or phrase with *sic*.

Finally, a less common abbreviation is viz., which stands for the Latin *videlicet*, which in turn is a contraction of *videre licet* (literally "it is permitted to see"). The use of viz. is very close to that of i.e.; it is used when introducing a list, but mainly when the list specifies a foregoing generality rather than merely restating or enumerating it. For example, "We had the essentials for field work, viz. binoculars, notebook, and pencil." Think of i.e. as replacing the English "that is" and viz. as replacing "namely."

Abbreviating scientific names

Every described species of organism has a Latinized scientific name, which may be abbreviated under certain circumstances. Each species has a unique binomial, the first name being that of the genus and the second the species. A genus may comprise several different species, and species in different genera may have the same specific name, so only the combination is unique. For example, the black-capped chickadee is *Poecile atricapilla* and the Carolina chickadee is *Poecile carolinensis*, whereas the white-breasted nuthatch is *Sitta carolinensis*. As shown, the Latinized binomial is italicized and the generic name begins with an initial capital. Subspecific designations are trinomials, the subspecific name appearing after that of the species and written similarly.

Two rules govern abbreviation of scientific names. The first is you may abbreviate a name when the one following it in the same binomial or trinomial is identical. For example, the black-crowned night-heron is *N. nycticorax* and the subspecies of the common snapping turtle over most of its range in the USA is *Chelydra s. serpentina*. A subspecies of the American bison is *B. b. bison*, the abbreviations of both generic and specific names allowed because they are both identical with the subspecies name. The name abbreviated must, however, be truly identical with that following it; be alert to names that are merely similar, as that of the common grackle, *Quiscalus quiscula*.

The second simple rule is that you may abbreviate a generic name whenever it is the same as the last mentioned genus beginning with the same letter. It is important to note, however, that this rule is subordinate to the first one. For example, one could write "the Japanese macaque (*Macaca mulatta*) and rhesus macaque (*M. fuscata*) are two species of Old World monkeys," and the abbreviated name would be read correctly as *Macaca* because "*M.*" could not stand for "*fuscata.*" You could not reverse the order to "*Macaca fuscata* and *M. mulatta*" because the abbreviated name would be misread as *Mulatta* by the first rule, instead of referring correctly to the actual genus name *Macaca*.

Sometimes a rewrite is necessary in order to avoid an ambiguity of abbreviation. For example, a study concerned three European species in the same genus: jackdaw (*Corvus monedula*), rook (*Corvus frugilegus*), and carrion crow (*Corvus corone*). A student of one of us quoting this study initially listed the species in the order of the original study as *Corvus monedula, C. frugilegus,* and *C. corone,* but that would not work because the genus of the latter species must be read by the rules of nomenclature as *Corone,* not *Corvus.* Writing out the genus for the third species without also writing it out for the middle one would be awkward and a tad confusing, and writing out all three would unnecessarily waste space. The solution was to reorder the list as *Corvus corone, C. monedula,* and *C. frugilegus,* thereby avoiding ambiguity while preserving parsimony.

Use common sense when abbreviating scientific names. The full generic name abbreviated should have occurred recently, and we offer this rule of thumb:

RULE

Abbreviate generic names only when the full name is in the same paragraph.

Not only is the reader bothered by having to thumb back several pages to identify a genus, but furthermore large separations of full name and abbreviation can introduce ambiguities to your writing. For example, if you mention a genus in one paragraph and abbreviate it in the next, you may not be aware of that separation when inserting a new paragraph between the two. If that new paragraph mentions a different genus beginning with the same letter, then the abbreviation in the following paragraph then refers to the wrong genus. We have seen such wrong references in published papers and surmise they came about thus.

Initial capitals in avian names

One of our anonymous referees asked about the use of initial capital letters in common names of species, so we add these notes. Throughout most of zoology and botany, the common names of organisms are traditionally written wholly in lower-case letters. Popular works on wildflowers sometimes use initial caps, but the only place this convention is applied consciously in scientific writing is in ornithology. Virtually every English-language ornithological journal in the world – perhaps hundreds of them, counting regional, state, and local journals (which are common in ornithology) – requires avian common names to be written with initial caps. There is a valid reason for that convention, which allows easy distinction between a "blue jay" (some jay that is blue in color) and a "Blue Jay" (the species *Cyanocitta cristata*).

Ornithology is special because common names have been legislated for a good bit of the planet by national ornithological societies. In fact, a study of US bird species found common names to be more stable than scientific names, thanks to this legislation by the American Ornithologists' Union. There has been an increasing trend in mammalogy and herpetology to standardize the common names of other terrestrial vertebrates, so we might witness in coming years a parallel adoption of initial capitalization for names of mammals, reptiles, and amphibians.

USING SIGNS, SYMBOLS, AND MARKS

A few abbreviated forms of writing trouble many authors in science. Most abbreviations used in scientific writing conform to well-known standards, such as SI for units of quantities, discussed earlier. The author must explain non-standard abbreviations that he or she employs, as in "The abbreviation 'LT' will refer to the loud threat call." Here are two kinds of abbreviated construction that authors frequently misuse. Both result from a keystroke on your computer that governs something other than a letter or numeral. To the discussions of these symbols (- and #), we add comment on the exclamation mark (!).

The hyphen (-) and the dash (–)

The hyphen (-, a short line) is used to make compound words. Hyphenation of text has been changing over the past several decades.

Formerly, hyphens were used extensively to create compound nouns, as in "unit-abbreviation" or "check-list." The modern tendency has been either to substitute a space ("unit abbreviation") or simply to eliminate the hyphen ("checklist"). Substituting a space can lead to strings of nouns, some of which are meant to be compounded and others intended to be modifiers without adjectival or adverbial inflections. Usually, the best solution in order to avoid ambiguity and confusion is to rewrite (see "Noun string construction grammar"). Another solution that has been emerging in English is to hyphenate pairs of nouns used as modifiers when adjectival and adverbial forms do not exist. Thus, "tab-delimited format" is a common expression in computing, there being no adverbial form of "tab" available to modify the adjective "delimited." Note that it is not correct, however, to insert a hyphen when grammatical forms exist, as in "sexually receptive female."

The en dash (–, a longer line) is used to express a numerical range. As typewriters did not have en-dash keys, the hyphen was used in place of the en dash. This usage continues today because authors rarely know how to use the en dash. Typesetters and editors will probably change hyphens to en dashes where necessary.

It is commonplace to separate the lowest and highest value of some variable to express its range of variation, as in "Chickadees have wing lengths of 59–68 mm." The abbreviated construction "59–68" stands for "from 59 to 68." Therefore, it is a pleonasm to write "from 59–68" for that literally means "from *from* 59 to 68." You might be tempted to consider that distinction trivial, but a reader who knows what the range hyphen really stands for will trip on "from 59–68" and the last thing a writer wants is to break the reader's concentration by trivia. Another way to express the true referent of a range hyphen is "between 59 and 68," so mistakenly writing "between 59–68" is equally redundant and confusing.

The en dash is also used to set off particularly long elements of a sentence, serving somewhat as a "super-comma." Dashes may be used in pairs internally in a sentence – as frequently employed in this

book when the material is too important to be relegated to a paren-
thetical expression but so long as to require being set off from the rest
of the sentence – or singly to set off a terminal expression.

The em dash (—, the longest of the three lines) is used as per the
en dash to separate elements of a sentence, but there are no spaces
either side of an en dash.

New uses are emerging for the hyphen and the dash in English,
and it is not certain what final convention will emerge. For example,
the first US states that amalgamated their institutions of higher
learning under one title modified by a city name used either a comma
or dash, as in "University of California, Berkeley" or "State University of New York – Stony Brook." Increasingly, these forms have
given way to the hyphen, as in "University of Wisconsin-Madison,"
which is probably the worst of the three because it compounds the
two hyphenated words instead of the entire first part with the entire
second part.

The number symbol (#)

The number symbol (#) is most frequently misused in labeling
figures, and attempts to use it in text are immediately changed by
copy-editors. Copy-editors see the number symbol as an informal
abbreviation for a word ("number") that should be spelled out, but
in fact this arbitrary sign does not stand for "number" in all
its senses. Consider this example: "The number-one cause of mortal-
ity was predation, and the number of frogs that died in a month's
period was eight." One may write "The #1 cause . . ." but not "the #
of frogs . . ." Still, it would be more graceful to write "The principal
cause . . ." even though "The #1 cause . . ." would also be correct.
Many cases exist in science, however, in which using the number
symbol to express rank order is useful: "This dove was #17 to arrive
at the drinking hole." As the number symbol is not a synonym for the
word "number," do not label an axis on a graph with something such
as "# of strikes/min," but rather use "no." as the abbreviation for the
word "number." The distinction is that between ordinal numbers

(expressing order) and cardinal numbers (expressing quantity): use "#" only with ordinal numbers.

Here are a couple of further tidbits about the troublesome number symbol. Insofar as we can determine, the world has roundly ignored the supposed formal name "octothorp" for this symbol. The term "pound sign," though, has gained some use in recent years, as pound as a unit of weight. Furthermore, "#" is used by printing personnel in some countries (including the USA) to indicate where a space should be inserted in copy.

The exclamation mark (!)

Oh! The rule is so simple: use the exclamation mark after an exclamation. Yet even the best of writers is tempted to lay emphasis on a sentence by placing the misused mark at its end. Emphasis is not what this punctuation mark is all about; it is used to warn the reader that the foregoing is not a sentence but rather an exclamation. No need to find a subject–predicate construction, dear reader, so don't let that exclamation trip you up. One of us had an undergraduate advisee who, when writing an mail message, always began "Hi!!! How are you?!!!!!"

For the sake of completeness, we mention that there is an arbitrary symbol in mathematics that looks exactly like the exclamation mark of grammar. Of course, the mathematical symbol means something different, namely factorial. Thus, 4! means $4 \times 3 \times 2 \times 1$, and has nothing to do with an exclamation. Nonetheless, it is a fact of mathematics that $0! = 1$, so we are tempted to assert that any process creating something out of nothing really does deserve to be exclaimed about. Wow!

DIFFICULT INFLECTIONS

English is so degenerate compared with most Indo-European languages that there remains only a vestige of inflection: changing the form of a word to show grammatical features such as person, number, case, mood, tense, and so on. Where inflection is used, however, the proper form will help make writing clearer. We select a few cases that seem especially to trouble many science writers.

Matching number of subject and verb

Everyone knows that a singular subject demands a singular verb and plural subjects demand plural verbs. Nevertheless, experienced writers still confuse readers with subjects and verbs that fail to match in number. The commonest mistake occurs when a prepositional phrase having a plural object follows a singular subject; the writer seems to lose track of his or her own sentence structure and writes a plural verb. Authors are especially prone to this mistake when a dependent clause or other aside intervenes. For example: "The actual count of gazelles, which can be observed with binoculars, exceed the total determined by census estimates." The verb should be "exceeds," referring to "count" (not "gazelles" or "binoculars").

Collective nouns, such as "number," are especially tricky because they may take either a singular or a plural verb, depending on the context. In general, these collective nouns demand the singular verb when preceded by the definite article and a plural verb when preceded by an indefinite article. For example: "The number of papers published annually is huge, and a number of them contain improper inflections." In that sentence, "The number" demands the singular verb ("is") and "a number" demands the plural verb ("contain").

By contrast, the word "data" is at root plural despite the fact that many writers today state "The data is consistent . . ." and similar structures. The use of "data" as a singular noun is now sufficiently widespread as to be acceptable by a majority of authorities, even though some of us trip over such usage. The singular "datum" is underused, many writers tending to substitute "data point," which not only is longer but also uses a noun to modify a noun.

"Noun string construction grammar"

Many adjectives are merely nouns changed in form to be modifiers, such that ignoring the inflection makes nouns seem to modify nouns.

Such obliteration of parts of speech often hinders understanding, especially by readers whose native language is not English. The trend to ignore adjectival inflections seems to have been spurred by the use of unhyphenated compound nouns, especially new technical terms of the modern era. We thus have terms like "computer science," when "computing science" would have been less confusing. The communicative distortion is increased when such a compound noun is used to modify another noun, as in "computer science center," which might be a scientific center for computers or a center for computing science.

In general, we recommend avoiding "noun string construction grammar," but common sense should prevail. Titles in particular should be brief, so they may have to include shortcuts. One of us once submitted a manuscript to *Nature* under the title "Coding of the gull chick's colour preference," which the editors changed to "Coding of the preference for colour by the chick of the gull." The published compromise was "Coding of the colour preference of the gull chick."

Adjectival degrees

"Good–better–best" is the example drummed into the heads of children to remind them that comparative and superlative inflections may be irregular. (Many children sensing the regularity create the non-words "gooder" and "goodest.") What the example fails to remind us is that the comparative needs to make the comparison explicit or clearly implied. Even accomplished authors occasionally fall into the trap of writing constructions like "analysis by Chi-square is better" without explaining what it is better than.

Those awful subjunctives

We have all learned about the subjunctive mood (= mode) of verbs, but try specifying which of these two inflections is correct: (1) "If the monkey *were* female, it would not respond" or (2) "If the monkey *was* female, it would not respond." In fact, both are grammatically correct, and they have different meanings. A rule governs this choice: it is unambiguous but still difficult to understand. The true subjunctive indicates something contrary to fact: "If the monkey were female, it would not respond, but it was male so responded vigorously." The conditional that is so similar in form requires the indicative mood: "If the monkey was female, it would not respond, but if it was male, it would respond vigorously." The key distinction is between (1) contrary to fact and (2) conditional relations.

Americans have so lost track of the distinction between subjunctive and conditional that most modern writers seem not to understand it, and much less be able to express it. Perhaps we (collectively) are losing the ability to make the distinction inflectionally and will have to employ circumlocution to make our intent clear. In any case, wondering or asking about something never (by traditional grammar) uses the subjunctive. We should write and say "I wondered whether age was encoded by the bird's color bands" (not "I wondered whether age *were* encoded by the bird's color bands").

PROBLEMATIC PAIRS

Choosing the precisely appropriate word, instead of one that will or might do lamely, is the sign of a good writer. In general, we authors prefer to be "tight and traditional" in formal writing, rather than "loose and faddish." Darwin remains eminently readable today because he wrote carefully and conservatively. Most of the pairs of words below are examples of those that are commonly confounded in scientific writing, but a few entail other kinds of problems.

Affect/effect. Yes, *affect* is a verb and *effect* a noun as most frequently employed, but (with different meanings) *affect* is also a noun and *effect* is also a verb. An example of the frequent sense is "Temperature may affect behavior, but the effect is often small." (*Affect* as a verb also means to simulate or copy, as in "She affected a British accent.") Psychologists use *affect* as a noun meaning a strong feeling or emotion. Finally, *effect* as a verb means to bring about, as in "A new theory may effect a dramatic change in science."

Each time you use either of these words, it is wise to pause and be certain that you have untangled them.

Among/between. As a first approximation, *among* refers to more than two, and *between* to just a pair. Accepted grammatical usage, however, goes beyond this easy distinction. Both words can be used to refer to more than two, but with different implications. We can say "The fruit dropped between the three monkeys," meaning that it did not fall on or by any individual. Saying "The fruit dropped among the three monkeys" means it fell within the group without specifying whether it fell on or by any individual. For other tricky uses, consult an advanced grammatical source or an explanatory dictionary.

And/or. Presumably no one confounds *and* and *or*, but some writers use the compound *and/or*, especially in legal and other special uses. In general writing, however, *and/or* lacks grace and seems inarticulate. In English, *or* is always inclusive, meaning "Either A or B, or both." If you want the exclusive sense in English, you must write it out: "either A or B, but not both." There is no prohibition against writing out the inclusive sense as well, but it is unnecessary except for special emphasis.

As/like. If a more subtle distinction graces English, it has escaped our notice. Grammarians in the 1950s were quick to criticize the advertising slogan "Winston tastes good like a cigarette should," because it uses *like* as a conjunction in place of *as*. Things have not changed much; the 1993 edition of the *American Heritage Dictionary* notes: "If one uses *like* as a conjunction in formal style, one incurs the risk of being accused of illiteracy or worse." *Like* is basically a preposition – and, with related meanings, an adjective, adverb, noun, transitive verb, or intransitive verb – but never properly a conjunction. We can say "Coots swim like ducks," meaning in the manner of ducks, but we cannot properly say "Coots swim like we expected." The test is simple: when a verb is expressed following the dubious word, *like* is not the word you want. In most cases, *as* is the natural conjunction, but sometimes *as if* is called for, as in "It looks as if the weather will hamper field work today."

As/since. Although its metaphoric extension meaning "in as much as" is widely regarded as acceptable, *since* originally referred only to time, and many editors insist that it be so restricted to prevent confusion. When you are tempted to use *since* metaphorically, try *as* or *because* instead.

Because of/due to. Many writers are aware of objections to *due to* being used in place of *because of*, but they misunderstand the subtle grammatical point at issue. They believe that *due to* can refer only to things owed, as in "His payment was due to me." It is not, however, the underlying sense of the words to which grammarians object. Rather, they object to the employment of *due to* as a compound preposition equivalent with *because of*. Everyone seems to agree that when *due* is used as an adjective, it is always perfectly acceptable, as in "The postponement of the field trip was due to rain." By contrast, it is not considered correct by conservative grammarians to write "The field trip was postponed due to rain," because this usage amalgamates *due* and *to* into a compound preposition. Nevertheless, *owing to* is universally accepted in this construction, and *due to* is remarkably similar.

Both . . . share. Faulty constructions employing the range dash (discussed early in this appendix), such as "between 59–68 mm," are not the only forms of unnecessary and therefore confusing redundancy in writing. Here is an example of another frequent pleonasm: "Both sexes share the habit of turning leaves to find food." The words *both* and *share* are redundant and so together cause the reader to trip. One can write "Both sexes have the habit . . ." or "The sexes share the habit . . ." or even "The sexes have the habit in common . . ." but one clear statement of inclusion surely suffices.

But/however. This is an unusually difficult pair of words. Acceptable usage has been changing steadily. Basically, *but* is a conjunction and *however* is an adverb, both of which can express contrast. (Both have other meanings as well.) This pair of sentences shows their similarity:

These guidelines are complicated, but they are valuable.

These guidelines are complicated; however, they are valuable.

Conservative grammarians have sometimes insisted that neither word (in its contrasting sense) can begin a sentence. However, many respectable writers use *however* in this position (as in this very sentence). But the use of *but* initially (as in this very sentence) is still considered wrong by almost all authorities on English. Conservative grammarians sometimes recommend *nevertheless* as the initial word for showing contrast, but this word does not always work well because of its restricted connotation ("in spite of").

Compare to/compare with. There are three senses of *compare*; one requires *to*, another *with*, and the third is more or less oblivious to the distinction. When describing resemblances, use *to*: "The brain may be compared to a computer." When examining two things for similarities or differences, use *with*: "Wallace's formulation of natural selection should be compared with Darwin's." Finally, when likening one thing with another, use *with* if you are a traditionalist, although *to* is considered equally acceptable: "Lorenz's contributions are to be compared with (to) those of Tinbergen." These distinctions are admittedly subtle, but then subtle distinction is a hallmark of the accomplished writer.

Complement/compliment. Both words are transitive verbs and both words are nouns. Although they have distinct meanings, they are confounded because they sound the same and are spelled similarly. *Complement* means to complete or bring to perfection (and as a noun, that which brings about such an end). *Compliment* means to praise or express a courtesy (and as a noun, the act of doing so).

Compose/comprise. Strictly, the whole comprises the parts whereas the parts compose the whole. You can also say that the whole is composed *of* the parts, and the parts are comprised *in* the whole. Thus, a pod is composed of kangaroos, and kangaroos compose a pod. Or, a pod comprises kangaroos, and kangaroos are comprised in a pod.

Although contemporary writers have frequently used *comprise* in the sense of *compose* ("a pod is comprised of kangaroos, and kangaroos comprise a pod"), many authorities on English continue to condemn such usage because it obliterates a useful distinction and thereby invites misunderstanding.

Concert/consort. Both words have other senses, but the confounding usually involves the constructions *in concert* versus *in consort* when the meanings are close and sometimes overlapping. *Concert* refers to togetherness of intent or action whereas *consort* refers to physical togetherness: "Three scientists edited the journal in consort and worked in concert to improve its quality."

Depredate/predate. You can predate a check, and Darwin predates Tinbergen, but predators do not "predate" their prey. What do predators do to their prey? A common fix to this problem has been to use the verb *depredate*, but many readers sensitive to words rightly object because depredate means to plunder or ravage. If you don't like your predators simply to kill or catch their prey, better write around this problem.

Different from/different than. Both are acceptable, although *different from* is usually a better choice and the distinction in usage is subtle. *Different than* is used more by Americans than Britons, and tends to be restricted to contrasts other than between objects or individuals: "J. S. Haldane is different from J. B. S. Haldane," but "Ethology is different than it was when Tinbergen wrote *Study of Instinct*." Thus, when there is an implied clause, *than* flags it. For example, "How different this seems from Panama" suggests that the comparison is simply with another place, whereas "How different this seems than Panama" implies a clause such as "the way it was back when I did field work here."

Farther/further. *Farther* has the literal meaning of distance; *further* is the metaphoric extension. Even if you look farther down this page, you will find nothing further on this topic. Nevertheless, the two words are used interchangeably by many writers, and with some justification. After all, we say something is *far* from the truth, so in the comparative why do we need to say it is *further* from the truth?

Fewer/less. *Fewer* refers to countable items, *less* to continuous variables: the fewer the entries on this list, the less you will have to worry about. Apparent exceptions involving *less than* are usually illusory. We say "less than five weeks ago" and "less than $500," even though weeks and dollars are countable, because the underlying concepts of time and value are continuous. Still, we do say idiomatically things such as "an abstract should be restricted to 25 words or less." (We could say "25 or fewer words," but even good writers often don't bother.)

First/firstly (second/secondly, and so on). Some people believe that *first* is not an adverb, and so they must write *firstly* to begin a series, whereas others believe there is no such word as *firstly*. Both viewpoints are wrong. You may begin a list of points with either (both are legitimate adverbs), but be consistent in the list: first, second, third or firstly, secondly, thirdly.

Its/it's. Even authors who know the difference sometimes miswrite one of these words for the other. The problem is that English uses the apostrophe for both the possessive and contractions. Thus, "Jack has gone" can be written "Jack's gone" and "The pen of (belonging to) Karen" can be written "Karen's pen." By extension, then, the apostrophe might seem equally at home in contractions and possessions involving *it*. Here is a way to remember the distinction: other possessive pronouns, such as *his* and *hers*, do not have the apostrophe, so neither does *its*. Contractions always demand the apostrophe. It's easy to remember the possessive *its*.

Ladder/latter. These are not true homophones but they sound sufficiently alike that you might write one when you intended the other. We know because one of us did just that when drafting the manuscript of this book.

Parameter/variable. The original meaning of *parameter* has been distorted by its passage into common parlance. In strictly mathematical terms, a parameter is a special kind of variable, namely one that is held constant for given uses of a functional relationship. The metaphoric extension as a general term for any variable that limits the extent of something is now widespread. The main danger lies in extending the metaphor so far that the writing seems pretentious, for example using *parameter* as a synonym for *characteristic*. A good writer would not say "Feathers are an important parameter of birds."

Principal/principle. The distinction is simply one of meaning, but even experienced writers occasionally confuse these two words. *Principal* is the adjective meaning main or chief (although also a noun in the sense of a school principal). *Principle* is the noun referring to a basic truth, standard, policy, and so on. Thus, "Even many scientists do not appreciate the principal implications of the Heisenberg uncertainty principle."

That/which. *Which* merely introduces added information, which can be omitted from the sentence without disrupting its structure (as in this very sentence). *That* introduces something that specifies meaning (as in this very sentence). The added information introduced by *which* is always set off in commas, whereas *that* never takes a comma. If you tend to use *which* in place of *that*, we urge you to go on a "which hunt" through your drafts. Nevertheless, many good writers (especially in the UK) use *which* in the defining sense of *that* and therefore do not set it off with a comma.

Therefore/thus. *Thus* means in the *manner* described. Its specific meaning has been damaged slightly by extension at the hands of less experienced writers to the point that *thus* has been used nearly synonymously with *therefore*. The latter word is more general, meaning because of the reasons set forth, but using *thus* in this way rarely causes the reader to stumble. Note that *thus* is an adverb, and so there is no need to use the modern abomination "thusly."

Whereas/while. *While*, like *since*, originally referred only to time, but it has become confusingly generalized in much modern writing. In most such cases where you are tempted to use *while* metaphorically, *whereas* will serve a little better.

USING A WORD PROCESSOR

Writing on a computer offers many advantages over old typewriters, and all of us need to master the provisions of the word processor being used in order to avail ourselves of these advantages. The use of the Find function to check references was explained in Chapter 3. Two other principal provisions are headers/footers and spell checkers.

Headers, footers, and line numbers

Almost all modern word processors allow you to put headers or footers on every page of the manuscript by typing them in just once. Furthermore, you can include the page number by typing a special symbol (different in different word processors) so that your document is paginated automatically. *It is an absolute requirement of a manuscript that the pages be numbered.* For the manuscript of a report, we recommend a right-aligned header with your last name, short article title, and page number; for example, "Lorenz, Instinct, 17." Page numbers at the upper right of the page make it easier for the reader to find pages; centered page numbers and those in footers are less convenient.

Most word processors also have special provisions. A useful function suppresses the header/footer on the first page of a document. This provision may be called "Title Page" or something similar. Reports, theses, and manuscripts are generally printed only on one side of the page, but there are times when both-sides printing may be desired, as in handouts for classroom use. Word processors usually allow separate headers for odd- and even-numbered pages, so you can right-align headers on the former and left-align them on the latter (as in this book), although even these word processors do not always allow separate printing of even- and odd-numbered pages to make a two-sided original document. Always verify and test your word processor's capabilities with a few pages before sending a large file to print.

Some journals require manuscripts to bear line numbers in the left margin. Some, but not all, word processors have a provision for automatic numbering of lines. If your word processor does not do this, try to convert your final document to the format of a different word processor that does (say, on a friend's computer). Lacking that, you may be able to prepare line numbers on a blank page using the same font size as used for your manuscript. Set the left margin narrower than that of your manuscript, number the lines, and print a copy. You can then photocopy this base page and use the sheets to print your final manuscript. It's not a great solution, but you may have no other option. That ploy works only if each page may have separate line numbers. Rarely, a journal may require that lines be numbered straight through the manuscript. Maybe submitting to a different journal is the only door open to you.

Spell checkers

There is nothing more aggravating for a reader (peer reviewer, editor, supervisor) than continually encountering misspelled words – and there is no excuse for such carelessness. Almost all modern word processors have spell checkers built into them, and all of these will search your entire document, flagging doubtfully spelled words. Some can also be set to beep at you while you are composing or flag a word with highlighting, so the doubtful word can be checked immediately. If your word processor lacks a spell checker, get a different word processor; you will not regret the investment. Learn to use your spell checker to your best advantage.

It is necessary to activate, update, and maintain in a meticulous manner your own user dictionary. One damnable word processor marketed would have you add words to its basic dictionary; most word processors laudably provide a separate user's dictionary into which you put your words. Basic spell checking dictionaries include only the most frequently used words of common parlance. Any academic's vocabulary will easily include perfectly good English words not listed in the word processor's basic dictionary. And in scientific writing, you will need to add many technical terms to your user's dictionary. The critical thing is to keep a standard large hard-copy dictionary available near your computer so that you can double-check the spelling of a new word before entering it in your user's dictionary.

We add a few words of caution. First, remember that spell checkers cannot identify homonyms such as "there" for "their." Nor can they find misspellings when the wrong version also is a word, as in typing "how" or "not" when "now" was intended. The scientist must learn to read proofs anyway so should not depend upon spell checkers as the sole means of catching errors in manuscripts. Finally, many word processors have grammar checkers, but reviews of these checkers have not been enthusiastic. After all, the miracle of the modern desktop computer cannot do *everything* for you.

Real dictionaries

Spell checkers are not real dictionaries because they do not provide the meanings of words, but computer dictionaries are available on the market. A particularly good one is the deluxe edition of the *American Heritage Dictionary*, which provides not only annotations and derivations (from Latin, Greek, Indo-European, and so on) but also usage notes for difficult words. If you do not have a true dictionary on your computer, keep a hard-copy dictionary nearby for ready reference whenever you are composing or checking text.

A thesaurus is another reference book that the serious writer should have at hand. There are computer versions available. A good thesaurus provides not only synonyms (words with overlapping meanings) but also related and contrasting words, as well as antonyms (words with opposite meanings, such as "wet" and "dry"). The term "thesaurus" is also used for dictionaries of special words, such as a medical dictionary or a dictionary of scientific terms. This kind of thesaurus for your field, if available, may also be a useful adjunct to your reference books.

Layout on a word processor

Word processors provide more flexibility than typewriters did, especially in the design of how things look on paper. Most typewriters merely allowed you to set margins and tabs, and to choose between single and double spacing; you were stuck with the size and appearance of the machine's typeface. Word processors offer so many possibilities that you need to spend some effort familiarizing yourself with the options available.

Spacing, margins, tabs, and justification

Most manuscripts (Chapter 3), including theses, are double-spaced, but grant proposals (Chapter 2) are usually single-spaced. Photo-ready abstracts submitted for a meeting presentation (Chapter 4) are also usually single-spaced. Word processors generally offer an in-between option of a half line of space between lines of print (so-called 1.5 spacing). Such intermediate spacing might be useful in laying out a table in a manuscript.

Typewriter-like margins are sufficient for most uses. A standard manuscript format has at least 1 inch (2.5 cm) of margin on all four sides surrounding your text. If a document is to be stapled along the left side, you may wish to have a larger margin there. Many word processors go further in allowing you to provide a binding gutter. For most purposes, this option makes no difference: you could either have a binding gutter and then the standard 1-inch margin on the left, or you could simply make a slightly larger left margin. However, if you craft a document to be printed on both sides of the page, then a binding gutter will be placed on the left side of odd-numbered pages and on the right side of even-numbered pages, so that your text always appears centered on every page after binding.

Typewriter tabs simply allowed you to skip to a new fixed place on the line to resume typing, but tabulation in word processors is more flexible. Generally, you have four types of tab to choose among: (1) left-aligned tabs are like those on a typewriter; (2) right-aligned tabs line up the right sides of your text, so that the beginnings are ragged if items in a tabbed column are of different length; (3) center-aligned tabs center the typed item, so that both left and right edges are ragged if items are of varying length; and (4) decimal-point alignment lines up the decimal points in numbers. Here is an example of all four types at work:

(1)	(2)	(3)	(4)
473.07	473.07	473.07	473.07
3.1416	3.1416	3.1416	3.1416
5280	5280	5280	5280
22	22	22	22
0.00001	0.00001	0.00001	0.00001

Option (4) is recommended for tables in scientific writing because it facilitates rapid and accurate comparison of values.

Finally, you can align the full line of text in four ways that are similar to the four provisions of tabulation. The first three (left-aligned, right-aligned, and centered) are the same as the first three types of tab, and the fourth (justified) creates alignment of both left and right edges. Word-processing software sometimes refers to these four as left-justified, right-justified, center-justified, and fully justified, respectively. Word processors generally justify ("fully justify") only complete lines, merely left-aligning ("left-justifying") the last line of a paragraph. Justification ("full justification") works by adjusting the spacing between words on the line. Here is an example of all four types:

(left-aligned)
Now is the time for all good people to come to the aid of their nation.
The quick brown fox jumped over the lethargic domestic canine.
A lot of years ago, our ancestors brought forth on North America
a new country.

(right-aligned)
Now is the time for all good people to come to the aid of their nation.
The quick brown fox jumped over the lethargic domestic canine.
A lot of years ago, our ancestors brought forth on North America
a new country.

(centered)

Now is the time for all good people to come to the aid of their nation. The quick brown fox jumped over the lethargic domestic canine. A lot of years ago, our ancestors brought forth on North America a new country.

(justified)

Now is the time for all good people to come to the aid of their nation. The quick brown fox jumped over the lethargic domestic canine. A lot of years ago, our ancestors brought forth on North America a new country.

You might think that justification ("full justification") would be the desirable option, mimicking as it does books and most magazines, which are set by Linotype or a similar process. It is true that such justification gives documents a professional look. Nevertheless, studies of reading have shown that left-alignment promotes faster reading and greater comprehension. The reason seems to be that if the right edges are also aligned, then readers make more mistakes scanning to the beginning of the next line. The ragged right edge of left-aligned print apparently helps the reader's eyes move from the end of one line to the beginning of the next without error.

Font sizes and types

The bewildering array of fonts available for use in word processors provides great flexibility in designing the look of your document, but that variety also entails potential pitfalls. We consider several issues in choosing a font to meet your needs.

First, there is what is generally called style. Most proposals (Chapter 2), manuscripts and theses (Chapter 3), and meeting abstracts (Chapter 4) will be prepared in plain text, with such styles as **boldface** and *italics* being used for special words and phrases (such as titles, scientific names of animals, and so on). Journals now differ concerning how to deal with style in a manuscript for publication.

The older convention was to underline words in type to indicate italic and to mark boldface by wavy underlining or some other specified indication. Many journals now prefer manuscript style to be prepared using word-processed styles of *italic* and **boldface**. Some styles that you will have occasion to use only rarely include outline, shadow, and ~~strike through~~.

The size of type is confusingly designated by two different systems. Most word processors designate point size, which refers to the heights of letters and the small spaces above and below them that separate letters on two successive lines. Therefore, 12-point is larger than 10-point size. Typewriter sizes were generally reckoned on the number of standard letters printed per inch, so that a larger number of letters meant smaller letters. Here is a comparison:

> This is 10-point, the typewriter equivalent being elite (12 letters/inch).

> This is 12-point, the typewriter equivalent being pica (10 letters/inch).

Be aware that computer fonts of a given point designation vary noticeably in actual size, with (for example) 10-point in one font being nearly as large as 12-point of a different font. Most documents such as proposals (Chapter 2) and reports (Chapter 3) will be prepared in 10- or 12-point size. Smaller point sizes are difficult to read and larger ones are too wasteful of space. Even a crowded 10-point font, such as Times (which mimics newspaper print), may be difficult to read easily.

Another issue in fonts is whether or not they are spaced proportionally. All letters of a typewriter require the same width, be they narrow like "i" and "l" or wide like "m" and "w." If the letter does not take the entire space allotted to the typewriter spacing, there is simply more leading and trailing space around such monospaced letters. Word processors can provide fonts whose characters take up only the required space for the particular letter typed, so that spacing between letters is more uniform in these proportionally spaced fonts. These fonts mimic typeset printing rather than typewriter copy. Here is a comparison:

`This is a monospaced font, similar to typewriter copy.`

This is a proportionally spaced font, similar to printed copy.

We know of no modern studies comparing ease of reading, but most people profess to find proportionally spaced fonts easier and quicker to read.

The last important aspect of fonts is whether they are serif or sans serif. A serif is a small stroke that finishes off a letter at the top and bottom. Computers usually provide a wide choice of the two types. Here is a comparison:

This is a sans serif (proportionally spaced) font.

This is a serif (also proportionally spaced) font.

Reading studies show that most people find serif fonts much easier to read. The serifs seem to help guide the eye, keeping it on the same line as the eye scans from left to right. Sans serif fonts, however, have a special use for graphics, where they prove easier to read, especially in vertical or oblique orientations (as on a map or the vertical axis of a graph).

A FINAL SUGGESTION

Many of us who make a serious effort to write well still occasionally commit grammatical and other errors or confuse words. You can train yourself to overcome such mistakes by keeping a log of them, adding to it each time you receive feedback from colleagues, advisors, journal editors, and the like. It seems that just the act of writing down an error a few times helps to reduce its future occurrence.

Appendix B **Ethics considerations**

Planning research
    Originality of ideas
    Collaborations
    Approvals and permissions

Proposing research
    Originality
    Feasibility
    Other funding
    Reviewing grants: confidentiality and conflicts of interest

Presenting research
    Honesty equals accuracy
    Coauthors
    Publishing non-redundantly
    Presentations in context

Presenting scientific credentials

Final comments

Integrity without knowledge is weak and useless, and knowledge without integrity is dangerous and dreadful.

Samuel Johnson (1709–84)

We have mentioned some of the most obvious ethics considerations
in the preceding chapters, but ethics is so important at every stage of
planning, proposing, and presenting science that we think the topic
deserves stand-alone coverage as well. Most professional societies
have their own ethics statements, and most scientific journals have
explicit requirements regarding the originality of the work submitted,
the consent of all coauthors, and the adherence to regulations
regarding animal and human subjects. All scientists should familiarize
themselves with the ethics statements of the societies most closely
aligned with their fields, and of the journals to which they will submit
their work. These ethics statements and requirements are updated
periodically, so you should always confirm that you have the most
current version of the guidelines on hand. Here, we offer a set of common
ethics considerations culled from our own experiences as independent
scientific investigators, collaborators on others' initiatives, and faculty
advisors of undergraduate and graduate students and of post-docs.

In organizing these considerations, we follow the same sequence as the chapters in this book. Many of the same ethics considerations apply at all levels of science, from planning, to proposing, to
presenting science, and, ultimately, to summarizing your qualifications as a scientist in your c.v. Perhaps the most critical of all is one's
honesty and integrity. The adage applies:

> **RULE**
> Honesty is still the best policy.

PLANNING RESEARCH

Just as good planning is a prerequisite for writing successful proposals
and presenting results effectively, considerations of ethics at the

planning phase of your research will help you to avoid unnecessary problems at subsequent stages.

Originality of ideas

A large part of planning scientific research involves the pursuit of new ideas that build on preexisting knowledge. But new ideas rarely if ever emerge in total isolation. More commonly, ideas emerge from exchanges with advisors, colleagues, and peers, or from an article or talk given by someone else that stimulates a new line of thinking. It is easy to acknowledge the sources of ideas in scientific proposals and publications because published references or personal communications linked to specific ideas can be cited in the text. Publications also include acknowledgments sections, where more general contributions can be mentioned. But as a research project moves from its initial planning phases to a full-fledged proposal, the sources of, or contributors to, the ideas can easily be lost, forgotten, or overlooked unintentionally.

People differ in how sensitive they are about sharing their ideas without credit. In an ideal world, reciprocity would reign, and helping a colleague with a good idea would be repaid with their equally valuable help later on. Concern about sharing ideas and, therefore, being scooped by a competitor leads some laboratories to impose no-talk zones in extreme cases. These groups tend to treat ideas as intellectual property, to be protected as vigorously as material or personal property. At the opposite extreme is the philosophy exemplified by one of our Ph.D. advisors, which is that if you are worried about your ideas being stolen, then you shouldn't be pursing an academic career because you should each have so many good ideas that no one could possibly steal them all. More commonly, people just feel bad if their intellectual contributions go unrecognized or ignored, and this may deter them from being helpful in the future. If there is ever any ambiguity in the source of ideas, then it may be worth asking or writing to the person in question. This is especially the case if a talk or article stimulates your thinking, because the speaker or author may already be working towards the same question and may be at a more advanced stage in the process than you. A grant proposal is unlikely to be approved if reviewers know that a similar study on a related topic is already in process. Contacting the source is also a good way to introduce the question of whether a more official collaboration might be appropriate.

In general, advisors and their advisees communicate frankly about ideas for thesis and dissertation research. Everyone knows that advisors usually play an important role here. Misunderstandings are more likely to arise between postdoctoral associates and their hosts. New Ph.D.s naturally feel relatively "independent" and expect a more collegial relationship with a postdoctoral sponsor than a major professor. The danger is that in this new role, the younger scientist will quickly absorb ideas from the new research environment and come unconsciously to believe the ideas originated with him- or herself.

TIP

Keep a dated bound scientific dairy in which you write down ideas and their sources. The record will later help remind you of ideas received from others so that the sources can be properly acknowledged. The diary might also be helpful to show at some later date that an idea was yours before any contact with someone who mistakenly believes you appropriated his or her idea.

Collaborations

If you look back into older literature in almost any discipline, you will find that single-authored papers were once commoner than they are today. For a variety of reasons, modern science demands much more collaboration among scientists than was formerly true. Working with other scientists, however, requires special attention to questions of who plays what role in the research.

Ideas are among the most difficult type of contribution to credit appropriately, especially when they arise in the context of one or more exchanges. Other kinds of contribution are easier to identify and attribute to a particular person, and a safe assumption in these cases makes a good rule of thumb:

RULE

A person is a collaborator (unless they explicitly defer) if you could not pursue your study without them or a resource they provide.

The help of a collaborator may involve actual data analyses or merely access to an unpublished dataset, study subjects, or a research opportunity at a field site or laboratory. Collaboration usually implies coauthorship on any publications and presentations that emerge from the work, unless decided explicitly to the contrary. Although you do not need to obtain the permission of someone you would like to acknowledge for their contributions in your work, you do need permission from collaborators whose contributions will be used.

Discuss your professional relationship with potential collaborators at the time you are planning your research. This will not only clarify mutual expectations for all parties involved but also facilitate your preparations in many tangible ways. For example, a collaborator may provide a letter stating their enthusiasm about the collaboration and the specifics on what they will contribute for use in your grant proposals. A full collaborator may also be able to help subsidize the part of the study on which they are collaborating and may be willing to read and comment on proposals in which they have a direct professional interest. Sometimes collaborators will decline coauthorship and the responsibility that implies and simply ask that you acknowledge them in whatever publications you produce. Either way, identifying appropriate collaborators during the planning phase of your research can facilitate your ability both to obtain funds and to conduct the research.

It may not always be obvious who appropriate collaborators might be. That is why it is a good idea to discuss your interests with faculty advisors and colleagues. Others might know of researchers working on similar topics or of new research initiatives that their colleagues have under way or that they've heard about at meetings or from others. Depending on the nature of your relationship to a potential collaborator, and how you learned of their work, it may be more appropriate to ask a third party, such as your advisor, to make initial contact for you. If you are a student about to contact someone senior in your field, then it is a good idea to copy your advisor on your communication so that he or she is aware of your contact and is prepared to field questions on your behalf.

Other instances exist in which identifying appropriate collaborations is more obvious. For example, you would never just show up in someone's laboratory expecting to be permitted to conduct research there. Similarly, it would be unimaginable and, in some people's views, unethical to show up at someone else's field site expecting to conduct your own research there. Inquiring about the possibility of pursuing your research in someone else's laboratory or field site is a mandatory step in planning research and has the potential to lead to fruitful future collaborative opportunities.

Approvals and permissions

Determining whether you will have access to a suitable laboratory or field site to conduct your research is something that needs to be done during the planning phase. There is no point in elaborating a proposal only to discover that you cannot do the research you propose for logistical reasons. As we noted in Chapter 2, it may take a long time to obtain official permissions such as those needed to conduct research in foreign countries. For ethical reasons, you should begin the process of inquiring about permissions at the time you are planning your research so that you don't end up proposing a study that proves to be unfeasible.

PROPOSING RESEARCH
Originality

Grant applications must be original in their content, and the expectation of reviewers is that they were written by the principal investigator or by the student with input and ultimate approval of the advisor who is serving as the principal investigator, such as on NSF Doctoral Dissertation Improvement Grants. Plagiarizing someone else's proposal is unethical and can have serious repercussions.

Feasibility

One of the criteria that reviewers consider is whether a proposed study is feasible in terms of time, adequate sample sizes, and the expertise of the investigator(s). We consider it unethical to propose a study if you suspect it may be unfeasible for these or other reasons at the time you propose it. For example, you may have a brilliant idea but know in advance that you will not be able to implement it as proposed because you will not have access to a sufficient number of study subjects. Although you may need to revise what you actually do if conditions change in unanticipated ways, you should never propose a study with the expectation that you will change a major aspect of it as soon as funding has been obtained. You should consult with your advisor and the funding source's program officer in the event of unanticipated changes in any major aspect of your proposed research to confirm that you still have permission to use the funds approved for one project on another.

Other funding

As we noted in Chapter 2, it is common to submit proposals for the same study to multiple funding agencies. Some funding agencies require you to list any current funding for the project and any pending or planned proposals. They may also require you to explain how these other funds will, or would if they are granted, affect your present request. All funding agencies expect to be notified if you have obtained other funding that overlaps with the request they have approved. Funding agencies vary in the degree to which they will permit you to use funds they have approved for other purposes related to the research, and it is best to find out in advance what is permissible. One of us had a student who received two grants that each covered the same travel and living expenses for a pilot field study, but his proposal for laboratory analyses was not allowed. One of the agencies prohibited its funds from being used for anything other than the field component, but the other agency was flexible in permitting the student to shift the redundant travel funds to cover the laboratory expenses. It is unethical to withhold information about your funding status from agencies that have agreed to provide research funds.

Reviewing grants: confidentiality and conflicts of interest

Reviewing grant proposals is a confidential process because the reviewer gains advance access to someone else's original ideas before the investigator has had an opportunity to pursue them. Reviewers are advised to treat grant proposals as confidential documents and are usually requested to destroy the proposals after filing their reviews. The contents of proposals are also treated as intellectual property. It is unethical to co-opt the ideas an applicant proposes for one's own research or even to talk about the ideas with others.

This can create a dilemma for a reviewer who was selected to evaluate a proposal because of their familiarity or prior work on a particular research topic. Reviewers should read the abstract of the proposal, but if they are also pursuing similar research they may declare a conflict of interest so as not to raise questions about the source of their own ideas. Other reasons for declaring a conflict of interest include a close relationship of any kind with the applicant. Our philosophy about whether to declare a conflict of interest is simple: When in doubt, declare, explain the issue(s), and let the program officer decide whether you are being too cautious.

PRESENTING RESEARCH
Honesty equals accuracy

Whether you are writing up your research for publication or preparing an abstract for a talk, it is unethical to deliberately misrepresent your results or any aspect of your study that might affect how others interpret your results. There are many cases in which researchers can misuse statistics, but one of the most common errors we have seen is how investigators represent their sample sizes. For example, you may have designed a study involving 12 subjects in each of two groups for comparative purposes but ended up for whatever reason with too few subjects in one of the groups for statistical analyses. Honest researchers explain the limits of their sample sizes and report their results as best they can. Dishonest researchers attempt to finesse their sample sizes and present invalid statistical analyses. Reviewers of manuscripts submitted for publication are likely to challenge the authors in their reviews, but listeners at talks may not notice the slide in which sample sizes are mentioned and are more likely to give the speaker the benefit of the doubt. Although you may escape public scrutiny, you are still misleading your audience and therefore misrepresenting your data. It is critical to be as honest and as accurate as you can, even if doing so takes some of the bang out of your conclusions. We think it is better to be known as cautious with data and conservative in your interpretations than to make bigger claims than your data permit.

Coauthors

As we mentioned in Chapter 3, customs differ about what the sequence of coauthors on publications or talks signifies. Usually, the first author has conducted most of the research and written most, or at least the first complete draft, of the paper. The most senior of the authors, usually listed last, is the person in whose laboratory the work was conducted and therefore responsible for facilitating, if not funding, the research. We cannot emphasize enough how important it is to establish an understanding about coauthorship in advance. This includes not only whose names will appear on the paper, but also the sequence of names and the expectations of what each author's contribution to the study was and their responsibility for the manuscript. This responsibility includes not only conceptualizing, writing, or reviewing and revising the manuscript, but also which of the authors will be identified as the corresponding author. In some fields the first author is generally the corresponding author, but in other fields the last author is always the corresponding author because it is with his or her research group that the project will continue.

All authors should be given sufficient time to review and make revisions on the manuscript that are acceptable to the other coauthors. All authors should understand that the contribution is original and has not been submitted elsewhere for publication, and all authors should agree with the accuracy of the data being reported. Some journals require that all authors sign a declaration to this effect, which may also serve as a transfer of copyright. Other journals expect the lead author to obtain the approval of all coauthors and accept the lead author's declaration as sufficient.

Publishing non-redundantly

Manuscripts submitted for publication are expected to represent original work that has not been submitted for publication or published elsewhere. Nonetheless, there is a growing trend in many fields to maximize the number of publications by breaking down research results into what is commonly known as their "least publishable units," or the smallest unit of data that will be accepted for publication. We have both noticed this trend by some authors in our fields, and although it is not technically unethical, it is usually obvious to anyone who is interested in a topic and therefore reads widely in it. Students should discuss their publication strategies with their advisors and colleagues. If you are working on multiple papers from the same database at the same time – as is often the case when writing a Ph. D. dissertation – then it is important to be able to distinguish what the major point or points of each publication will be. If you can't state in one sentence what distinguishes two papers from one another, then it is unlikely that your readers will be able to do so either.

All students learn early on that it is unethical (and illegal) to plagiarize. When referring to someone else's published work, the author or authors should be referenced accordingly. Direct quotes should be identified as such, with the appropriate citation and page number from which the quote was taken.

A similar approach should be taken when you are referring to your own previously published work, even if such citation appears to be self-promotional. Readers need to know whether you have described the same methods previously or referred to the same findings that you are presenting in your current paper, and they need to be able to refer to that source if they are interested. It is more difficult for new investigators submitting multiple manuscripts at the same time because there may not yet be an accessible source to cite. In this case, we recommend selecting one of the papers for the more elaborate treatment and referring reviewers and readers to it. If the second paper is accepted for publication before the first, then you may need to fill in the details and reverse the referencing between the two papers. However you manage it, the important point is to avoid plagiarizing yourself.

The last point invites further elaboration. Unethical authors, in order to increase their bibliographies, have been known to lift whole passages, and even major sections, from their previous publications as a mechanism for avoiding the tedium of new exposition. The phenomenon is called autoplagiarizing and is definitely something to avoid.

Another kind of plagiarizing is publishing essentially the same paper in two different languages. This habit arose quite innocently in earlier decades before English became the universal language for scientific publication. Researchers in countries where the major language was not English had to publish in their native language in order to get ahead academically. By the same token, they also had to publish in English to be widely read internationally. Fortunately, this almost mandatory dualism became unnecessary as major journals in every county allowed, and in many cases encouraged, publication in English. Therefore, what once was innocent, and even desirable, duplication of papers in different languages would today be considered unethical.

Finally, another obvious form of plagiarism is translating a paper and publishing it as your own. This kind of thing happened prominently a few decades ago in a biological paper, and some classic cases are renowned. Russian mathematician Nikolai Ivanovich

Lobachevski (1792–1856) is perhaps better remembered for his republication of the work of others without attribution than for his own contributions to hyperbolic geometry.

Presentations in context

Talks and seminars present a more limited context in which to refer to previous work by yourself or others. Unlike a manuscript, there is no place in a talk to provide a complete list of references cited, and when an author refers to published work the slide is usually limited to the author's name and year of publication. It is important, nonetheless, to provide accurate contextual information in a talk and to give credit to the original source of the idea in this way. It is inappropriate, if not unethical, to describe hypotheses or predictions that were generated by others as if you had derived them yourself.

PRESENTING SCIENTIFIC CREDENTIALS

Most reviewers will assume that an author was careless instead of dishonest about misrepresenting results in a manuscript, but the same cannot be said for reviewers of a c.v. Here, discrepancies, inaccuracies, or inflated representations of one's scientific credentials will be readily interpreted as deliberate, thus tarnishing your reputation and raising doubts about your overall integrity as a scientist. It is, therefore, important to take care in how you list achievements on your c.v. and how you describe them in a cover letter accompanying your c.v.

Errors and inaccuracies can creep into a c.v. unintentionally, and it may therefore be helpful to ask your advisor or a senior colleague to review a draft of your c.v. before you send it around. One of our students initially listed the grant that her postdoctoral sponsor had obtained as if it were her own. It was an understandable mistake, because it was through this grant that her postdoctoral fellowship was funded; nonetheless, it was misleading, because the way it was listed implied that the student, instead of her sponsor, had successfully secured the funding. Another of our students, by contrast, neglected to list an important source of funding that she had independently obtained, an omission that her advisor, but no one else, was quick to catch.

FINAL COMMENTS

We conclude this appendix with a reminder of our rule of thumb about the importance of honesty and integrity at all stages of the scientific process. The most valuable attribute any scientist possesses is his or her credibility. Allegations of scientific misconduct involving leading researchers make headlines in the news because they are so serious if found to be substantiated. Once your reputation is lost, it can be difficult, if not impossible, to recover it. Don't let the pressures to secure funding, publish, and promote yourself corrupt your scientific integrity. Many honest mistakes can be caught and corrected before they damage your own and your collaborators' reputations, but this requires constant vigilance and self-awareness and sufficient lead time during each phase of the scientific process.

Further reading

Alley, M. (1996). *The Craft of Scientific Writing*, 3rd edn. New York: Springer. [Examples of strong and weak scientific writing and suggestions for improving one's own.]

Alley, M. (2003). *The Craft of Scientific Presentations: Critical Steps to Succeed and Critical Errors to Avoid*. New York: Springer. [Presentations.]

American Psychological Association (2001). *Publication Manual of the APA*, 5th edn. Washington, DC: APA. [A style manual followed not only by the many periodicals of the APA itself but also by several other journals. More than half a million copies in print.]

Anholt, R. R. H. (1994). *Dazzle 'Em with Style: The Art of Oral Scientific Presentations*. New York: W. H. Freeman and Co. [Communicating and lecturing.]

Berg, K. E. and R. W. Latin (1994). *Essentials of Modern Research Methods in Health, Physical Education, and Recreation*. Englewood Cliffs, NJ: Prentice Hall. [Research and methodology in special biological areas, and technical writing.]

Beveridge, W. I. B. (2005). *The Art of Scientific Investigation*. Caldwell, NJ: Blackburn Press. [Originally published in 1950, a time-honored work on doing science, with many good examples of inductive reasoning.]

Beveridge, W. I. B. (1980). *Seeds of Discovery: A Sequel to The Art of Scientific Investigation*. New York: Norton. [More of Beveridge's germane analyses; see previous reference.]

Bishop, W. (1990). *Something Old, Something New: College Writing Teachers and Classroom Change*. Carbondale, IL: Southern Illinois University Press. [Writing and rhetoric in education.]

Booth, V. (1993). *Communicating in Science: Writing a Scientific Paper and Speaking at Scientific Meetings*, 2nd edn. Cambridge and New York: Cambridge University Press [Communicating, lecturing, and writing.]

Brent, D. (1992). *Reading as Rhetorical Invention: Knowledge, Persuasion, and the Teaching of Research-based Writing*. Urbana, IL: National Council of Teachers of English. [Rhetoric and reading.]

Briscoe, M. H. (1995). *Preparing Scientific Illustrations: A Guide to Better Posters, Presentations, and Publications*, 2nd edn. New York: Springer-Verlag. [Visual presentations and publications.]

Carr, J. J. (1992). *The Art of Science: A Practical Guide to Experiments, Observations, and Handling Data*. San Diego, CA: HighText Publications. [Experimentation, methodology, laboratory.]

Chamberlin, T. C. (1965). The method of multiple working hypotheses. *Science* **148**:754–9. [A reprint of Thomas C. Chamberlin's 1890 article from *Science* (old series) in which he artfully describes how to consider alternative hypotheses without bias.]

Cleveland, W. S. (1994). *Elements of Graphing Data*. Murray Hill, NJ: AT&T Bell Laboratories. [A standard comprehensive treatise on graphing.]

Cone, J. D. and S. L. Foster (1993). *Dissertations and Theses from Start to Finish: Psychology and Related Fields*. Washington, DC: American Psychological Association. [Practical suggestions for psychology graduate students.]

Cooper, H. and L. V. Hedges (eds.) (1994). *The Handbook of Research Synthesis*. New York: Russell Sage Foundation. [Storage and retrieval systems, general and statistical methods.]

Council of Biology Editors (1994). *Scientific Style and Format: The CBE Manual for Authors, Editors, and Publishers*, 6th edn. New York: Cambridge University Press. [A standard guide for writing in biology.]

Davis, M. (2004). *Scientific Papers and Presentations*. New York: Academic Press. [Writing and presenting science.]

Dawkins, M. S. and M. Gosling (eds.) (1992). *Ethics in Research on Animal Behaviour*. London: Academic Press for the Association for the Study of Animal Behaviour and the Animal Behavior Society. [Reprintings of eight articles originally appearing in the journal *Animal Behaviour*, including the revised guidelines for the use of animals in research, with an introduction by the editors.]

Day, R. A. (1998). *How to Write and Publish a Scientific Paper*, 5th edn. Phoenix, AZ: Oryx Press. [Writing and publishing in scientific journals; a widely used reference.]

Day, R. A. (1995). *Scientific English: A Guide for Scientists and Other Professionals*, 2nd edn. Phoenix, AZ: Oryx Press. [Guide to technical writing.]

Friedland, A. and C. Folt (2000). *Writing Successful Science Proposals*. New Haven, CT: Yale University Press. [Proposal writing.]

Gastel, B. (1983). *Presenting Science to the Public*. Philadelphia, PA: ISI Press. [Treats a related topic beyond the bounds of this book; a very complete guide.]

Goldstein, M. and I. Goldstein (1984). *The Experience of Science*. New York: Plenum Press. [Discusses hypotheses, precision, accuracy, and other relevant topics for scientists.]

Gutavii, B. (2003). *How to Write and Illustrate a Scientific Paper*. New York: Cambridge University Press. [Writing and, especially, presenting figures.]

Halliday, M. A. K. and J. R. Martin (1993). *Writing Science: Literacy and Discursive Power*. Pittsburgh, PA: University of Pittsburgh Press. [Communicating and technical writing.]

Hirsch, T. J. (1992). *Working Research: Strategies for Inquiry*. Englewood Cliffs, NJ: Prentice Hall [Problem-solving, research methodology, and technical writing.]

Hopkins, R. A. (1974). *The International (SI) Metric System and How It Works*, 2nd edn. Tarzana, CA: Polymetric Services. [A book of nearly 300 pages on the SI system of numbers and units.]

Kuhn, T. S. (1996). *The Structure of Scientific Revolutions*, 3rd edn. Chicago, IL: University of Chicago Press. [A famous book, the first edition of which defended more strongly than did the revision the thesis that science advances mainly through discarding old models for wholly new ones.]

Leather, S. R. (1966). The case for the passive voice. *Nature* **381**:467. [Objections to blanket advice for always writing in the active voice.]

Lehner, P. N. (1998). *Handbook of Ethological Methods*, 2nd edn. New York: Cambridge University Press. [Classic descriptions of ethological sampling methods.]

Lemke, J. L. (1990). *Talking Science: Language, Learning, and Values*. Norwood, NJ: Ablex Publishing Corp. [Language and languages in education and science.]

Lindemann, E. (2001). *A Rhetoric for Writing Teachers*, 4th edn. New York: Oxford University Press. [Writing English.]

Lobban, C. S. and M. Schefter (1992). *Successful Lab Reports: A Manual for Science Students*. New York: Cambridge University Press. [Laboratory report writing.]

Mahaffey, R. R. (1990). *LIMS: Applied Information Technology for the Laboratory*. New York: Van Nostrand Reinhold. [Laboratory management, data processing, and information systems.]

Martin, P. R. and P. Bateson (2007). *Measuring Behaviour: An Introductory Guide*, 3rd edn. Cambridge and New York: Cambridge University Press. [Methodology in studies of animal behavior.]

Montgomery, S. L. (2002). *The Chicago Guide to Communicating Science (Chicago Guides to Writing, Editing, and Publishing)*. Chicago, IL: University of Chicago Press. [Writing and presentations; also includes communicating with the public.]

O'Connor, M. and F. P. Woodford (1975). *Writing Scientific Papers in English: An ELSE-Ciba Foundation Guide for Authors*. The Hague: Ciba Foundation. [A guide conceived for writers whose first language is not English but very useful to native English speakers. Reprinted in 1977 by Elsevier/Excerpta Medica/North Holland.]

Ogden, T. E. and I. A. Goldberg (2002). *Research Proposals: A Guide to Success*, 3rd edn. New York: Academic Press. [Proposal writing and review process.]

Platt, J. R. (1964). Strong inference. *Science* **146**:347–53. [An influential paper urging that the hypothetico-deductive method be applied more explicitly in all branches of science.]

Secor, M. and D. Charney (eds.) (1992). *Constructing Rhetorical Education*. Carbondale, IL: Southern Illinois University Press. [Language and teaching.]

Smith, R. C., W. M. Reid, and A. E. Luchsinger (1980). *Smith's Guide to the Literature of the Life Sciences*, 9th edn. Minneapolis, MN: Burgess Publishing Co. [Although outdated and apparently no longer periodically revised, this remains a useful entry into biological literature searching and has special chapters on graduate theses.]

Sternberg, D. (1981). *How to Complete and Survive a Doctoral Dissertation*. New York: St Martin's Press. [Includes tips on many unexpected problems that graduate students may encounter.]

Strunk, W. and E. B. White (2000). *The Elements of Style*, 4th edn. New York: Longman. [The parsimonious classic on grammar and usage, comprising the lecture notes of Professor William Strunk (1869–1946) as originally edited by his once-student, the US essayist E. B. White (1899–1985), who added an introduction and a chapter on writing. This edition includes a foreword by R. Angell.]

Swales, J. (1990). *Genre Analysis: English in Academic and Research Settings*. Cambridge: Cambridge University Press. [How language can be used in research and teaching. One reviewer of our book in manuscript especially recommended the way this author discusses the purpose of a paper's introduction.]

Tufte, E. R. (2001). *The Visual Display of Quantitative Information*, 2nd edn. Cheshire, CT Graphics Press. [One of several works by Tufte that have become standard reference material; see next reference.]

Tufte, E. R. (2003). *Envisioning Information*. Cheshire, CT: Graphics Press. [A concise guide to presenting data visually, mainly with reference to computer interfacing; the 2003 edition is its ninth edition.]

University of Chicago Press (1987). *Chicago Guide to Preparing Electronic Manuscripts for Authors and Publishers*. Chicago, IL: University of Chicago Press. [Now outdated because of the rapid advances in microcomputer hardware and software, but most of the basic material remains valid.]

University of Chicago Press (2003). *The Chicago Manual of Style*, 15th edn. Chicago, IL: University of Chicago Press. [Arguably the best-known standard of style manuals since the first edition in 1906.]

US Government Printing Office (2000). *Style Manual*. Washington, DC: US Government Printing Office. [The standard manual for writing by federal employees, updated at long intervals; especially known for its detailed treatment of syllabification and hyphenation, which in the days of typewriters was a critical issue for breaking up words at the ends of lines.]

Watson, J. D. (1993). Succeeding in science: some rules of thumb. *Science* **261**:1812–13. [Advice from a Nobel laureate, who says to avoid dumb people, take risks, have someone back you up, and never do anything that bores you. See also the news article on grantsmanship in the pages preceding Watson's article.]

Zimmerman, D. E. and M. L. Muraski (1995). *The Elements of Information Gathering: A Guide for Technical Communicators, Scientists, and Engineers*. Phoenix, AZ: Oryx Press. [Communicating, technical information, storage and retrieval systems.]

Index